CROWS, JAYS, RAVENS AND THEIR RELATIVES

CROWS, JAYS, RAVENS
and their relatives

Sylvia Bruce Wilmore

DAVID & CHARLES
Newton Abbot   London   Vancouver
PAUL S. ERIKSSON
Middlebury, Vermont

ISBN 0 7153 7428 1 (Great Britain)
ISBN 0-8397-1894-2 (United States)
Library of Congress Catalog Card Number: 77-79245

Set in 11 on 13pt Ehrhardt
by Ronset Limited, Darwen, Lancashire
and printed in Great Britain
by Biddles Limited, Guildford, Surrey
for David & Charles (Publishers) Limited
Brunel House   Newton Abbot   Devon

Published in the United States of America
by Paul S. Eriksson, Middlebury, Vermont

Published in Canada
by Douglas David & Charles Limited
1875 Welch Street   North Vancouver   BC

CONTENTS

Chapter 1

DISCOVERING THE CROW FAMILY

The Crows, Jays, Pies, Nutcrackers and Choughs belong to the *Corvidae* family and occur in most parts of the world. They are the largest of the Passerine birds, with conspicuous personalities and great sagacity. Bold, active, noisy and aggressive, some of them brilliantly coloured, throughout the ages they have claimed man's attention; no other group except possibly the Owls, has earned a more prominent place in legend and folklore.

The family is of considerable age, its closest relatives being the specialised Orioles, Bowerbirds, Birds of Paradise, Shrikes and Drongos.

Linnaeus, the Swedish naturalist, listed the *Corvidae* family in the order *Passeriformes* in his 'Systema Naturae' in 1758. Since then the family has either been placed first or last in the Suborder *Passeres*. Its members are classed among the 'Higher Birds' because they have a relatively high mental capacity.

The twenty-six genera belonging to the family include over a hundred species. The groups vary a good deal in structure and habit and it is difficult to assign several intermediate genera to particular subfamilies. But all the species share important structural features as well as behaviour patterns.

The sombre colours of black and grey are generally associated with Crows, but the family contains some exquisitely coloured birds such as the Azure-winged Magpie, the Blue Jay, the Green Magpie and the Green Jay. Nutcrackers are distinctive with their chocolate and white plumage. Even the Crows and Ravens have variations of plumage.

The colours black, grey, brown and tan which are found in the *Corvidae* plumage are produced by pigments known as melanins; these occur as separate, distinct particles of definite shape and measurable size and are created in highly specialised, branched cells known as melanocytes. The non-iridescent blue plumage of such birds as the Blue Jay contains no pigment; the colour is a purely structural effect dependent upon the fine, colourless framework of feather barbs. The narrow black transverse bars on the outer surface of the vanes of blue-plumaged birds such as the Jay, *Garrulus glandarius*, appear to be the result of localised defects in the structure of the box cells of the cortical layer of barbs.

The brilliant green plumage of the Green Magpie of the Himalayas is produced in the same way as the blue, although the reflective part of the barb may be pigmented by substances called corontenoids which are carried by body fluids into the living cells of the feathers.

Many *Corvidae* plumages have an iridescent effect. The Magpie, *Pica pica*, has a boldly contrasting iridescent black and white plumage. In good light the Raven's black plumage can be seen to have an iridescence, and Crows have a greenish gloss rather than the reddish-purple of Rooks. Iridescent colours are produced by broad, flat feather barbles with blunt ends. The broad surfaces often overlap one another like the horizontal laths of a closed venetian blind, giving the feathers a smooth, satin-like surface. There may be various colours which are caused by the interference of light waves reflected from the outer surfaces of the barbules. This is the type of coloration seen in oil films on water and roads. Most *Corvidae* plumage looks shiny because the feathers lie with their flat reflective surfaces exposed.

The wings of the *Corvidae* have ten distinct primary feathers, the first exceeding half the length of the second, and the inner webs of the four outermost primaries having waved edges. An eleventh primary, 1cm in length, is termed a 'remicle'. The *Corvidae* tail has twelve feathers. The legs and feet are large and strong, the tarsus generally smoothly scaled with the back terminating in a ridge. The long-clawed feet are suitable for all purposes – perching, walking and scratching. The tough bills are longer than their depth, being generally little shorter than the head. In some cases the upper mandible has a simple notch. The nostrils are clear of the line of the forehead and are usually hidden by forward-growing bristles.

Corvids inhabit forests, bushland and grassland; mostly omnivorous and seldom finicky over their choice of food, they have been able to survive changing conditions in a variety of habitats. Some groups have a strangely limited distribution, and other species have widely disrupted ranges.

Most species feed on the ground but a few are arboreal. They are known for their habits of burying and hiding food and breaking nuts and shells with their bills while holding them with their feet. Many of them are predatory feeders and eat carrion. In common with all perching birds they have an aegithognathous palate, ie it is cleft down most of the bony palate but the vomers are forked at the front instead of coming to a point. There are taste buds on the base of their non-tubular tongue and on the palate. Corvids do not have a greatly developed sense of taste but it is thought they can distinguish between the four primary taste sensations, sweet, bitter, sour and salty. Their sense of smell is not very great either, the nasal chambers being directly connected to the mouth and nostrils, which enables them to smell food while in the mouth, and allows carrion eaters and birds of prey to spit out tainted meat. A gamekeeper has seen a Crow pick up poisoned meat and throw it into the sea.

Mating among the *Corvidae* entails a long and continued courtship. The sexes look alike; they have cloacal intercourse, and are mostly monogamous. Their family life is exceptionally united and protective. They are nearly all single brooded and the female usually broods and incubates.

The mean clutch-size of most Crows, and possibly Jays, increases with the degrees of latitude, but Crows reduce their clutch-size on small islands. The Corvid reproduction rate is not abnormally high.

Non-colonial breeders such as Ravens, Crows, Magpies and Jays have longer laying seasons than the colonial species, Rooks and Jackdaws. Most incubation begins before the clutch is completed so that the eggs hatch at intervals. This ensures that if food is scarce the last chicks to hatch are the smallest and weakest and quickly starve, thus reducing the brood to the number which parents can effectively feed with most chance of success. Crows, Rooks, Magpies and possibly Jays eat eggs that fail to hatch. Both the Carrion and Hooded Crows remove dead young of all sizes; Magpies remove very small nestlings; Ravens, Rooks, Jackdaws and Jays do not remove large nestlings, and Choughs do not remove their dead at all. With Carrion Crows and Rooks nestling deaths occur in the first third of the nestling period, whereas the nestling deaths of Jackdaws, Magpies and Jays tend to occur in the second and last third of the nestling period.

Crows and Rooks brood their nestlings for longer than Magpies and Jackdaws; they also search further afield for food. The female's brood patches gradually lose their down; at the same time they become swollen and covered with blood vessels which effectively conduct heat to the eggs as they come in close contact with the bird's body. Very young Rook and Crow nestlings require frequent feeds and suffer a higher mortality than other species.

Man destroys a high proportion of Corvine nests and is res-

1 Parts of the *Corvidae* anatomy (facing page)
A Sternum of adult Raven *Corvus corax* from below showing muscle attachment (dotted)
cb costal border of sternum
ps pectoralis secundus
pt pectoralis tertius
lx lateral xiphoid
pm pectoralis major
mx median xiphoid
B Under view of the skull of a Raven
vo vomer
mp maxillo-palatine process
pa palatine
ptg pterygoid
q quandrate
bsph basisphenoid
sphr sphenoidal rostrum
C Syrinx of Magpie showing the diacromyodian attachment of the intrinsic muscles at the ends of the bronchial semi-rings
left side view
right dorsal view
Membraneous semi-rings (dotted)
Ti internal tympaniform membrane
st muscle from the side of the trachea to the upper end of the clavicle
1–6 syringial muscles, the 7th is hidden by the 6th; the 4th is hidden
D Palate of Raven (Aegithognathous)
mxp palatal process of maxilla
pt pterygoid
q quadrate

A

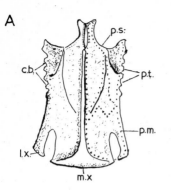

B

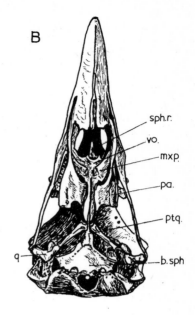

C

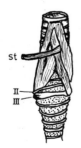

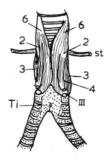

D

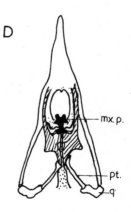

ponsible for many eggs failing to hatch. Corvids are predators of each other. Jackdaws raid Ravens' nests, Hooded Crows rob Magpies' nests, and Carrion Crows and Magpies rob Jackdaws' nests. Squirrels account for most of the mammalian predation and they are particularly harmful to Magpies and Jays. Weasels have also been known to steal Jackdaws' eggs. Rooks build their nests at the top of trees, and so, mainly, do Crows. Few predators, except man who organises Rook shoots, can destroy these birds' eggs, but the nests are so exposed that they are vulnerable to the weather.

Corvidae nestlings emerge from their shells poorly developed in every way except for their digestive system which requires constant feeding, and they are incapable of fending for themselves. Corvid nestlings belong to the nidicolous group of birds, which have the highest brain development, the indexes of their brain hemispheres being over ten. They have shorter incubation periods than nidifugous chicks which, though more developed at birth, nevertheless grow more slowly than nestlings and therefore require less food.

Newly hatched nestlings adopt the temperature of their surroundings, and it takes them from one to three weeks before they can maintain a constant temperature; so while they are in the nest they need almost continuous warmth from their mother. Therefore the female in many Corvid species remains with her young and depends on the male for food during most of the time the nestlings are with her. The parents feed the nestlings by regurgitating food from their food pouches; Crows bring water in their pouches to their young, or sometimes provide moisture by soaking their breast plumage. In Crows, the oesophagus, a passage from the pharynx through the crop (food pouch) to the stomach, has extensive folding to allow the storage of large food. The pouch is a storage organ whose function is to supply food as constantly as possible to the gastric apparatus where it is then softened and swollen. Crow nestlings gape to stimulate their parents to provide them with an almost constant flow of food; as they open their beaks wide, the bright pink interiors act as a stimulus for their parents to keep feeding them. In Crow nestlings the front of their pterygoid (a paired bone of the upper mandible) shows reptilian features, but later this becomes detached from the rest and fuses with the palatine, which forms part of the roof of the mouth. A movable joint is established between this joint and the main pterygoid. The pterygoid gives the impression of being separated from the vomers in the adult Crow but is not really so.

Until the seventh day of embryonic life the brain of a bird has a typically reptilian structure, but from the seventh to twelfth day marked avian characters appear. The brain of birds and mammals consists of three dilated regions, commonly known as the forebrain, midbrain and hindbrain. The cerebrum or cerebral hemispheres in the forebrain are the dominating co-ordinating centres.

Evolution of the avian brain to higher degrees of complexity is always combined with a tendency for the mass of higher centres to be concentrated in the front part of the hemisphere. In Corvids the frontal development is concentrated in the dorsal part of the sagittal elevation. The primitive arrangement of the olfactory bulb that controls the sense of smell is preserved but there is a marked decrease in its size in all *Passeres*, although this is far from uniform. As in birds such as the Corvids with their higher evolutionary levels of the hemispheres, the two bulbs may be partially or completely fused in one unpaired structure, as in the Magpie. The Corvids have reached high hemispheric indexes, with the Raven, *Corvus corax*, having indexes of eighteen to nineteen, whereas less highly evolved birds never attain indexes higher than 5.5.

Both Ravens and Jackdaws trained by Otto Koehler in Germany in the late 1940s could count up to six or seven; and research teams in America, Germany and Russia have reached the conclusion that many birds, especially the Crow family, tend to be quicker on the uptake than 'higher mammals', such as cats, dogs, monkeys etc. My tame Crow dominated my cat and dog, he would even pick up the spaniel's leash and lead him round the garden! Scientists at the University of Mississippi have been successful in getting the co-operation of Crows. One Crow learned to watch a clock face, wait until the hands pointed to twelve, and then operate a lever to obtain a reward of grain.

The ability to read a clock proves that Corvids have keen eyesight and an ability to concentrate. In birds the optic lobes are forced into a ventral position between the eyes and the labyrinth by reason of the mass of hemispheres and the cerebellum filling the narrow cranial cavity. The optic lobes of birds are not only relatively bigger than those of reptiles and mammals, their structure is also more complicated. The Corvid's eyes are placed to the front near the base of the beak, but although they are carnivores they are not specialised for this purpose as they neither possess the enormous eyes, nor the very long wings and the elongation of the forearm and hand, which enable birds of prey to soar and hover over their victims. Tests carried out to discover how efficient Crows are in seeking food proved that they have very keen eyesight, an ability to distinguish shapes and colours, and if presented with three different types of prey they can search for them all at once.

From all available sources it has been found that a bird's vision is not sharper, but a great deal faster, than that of man. They can assess the elevation of the sun and also its rate of change of elevation and of azimuth with great accuracy, but unlike man's, a bird's picture is flat and information about relative distances has to be built up from a succession of glances from different places.

Birds are eye-dominated and eye-dependent to a greater extent than any mammal, even man. Contrary to folklore, which claims that Ravens peck out living eyes, usually of their victims, Konrad Lorenz in his book, *King Solomon's Ring*, comments on an old German proverb which says, 'One crow will not pick out the eye of another.' Lorenz has proof of the Corvid's inhibition against pecking at eyes, when he held his pet Raven, Roah, on his arm and turned his face towards him with his eye near the bird's beak, and Roah withdrew his beak immediately. This inhibition could be an explanation of the effectiveness of eyespot patterns on insects as a defence against avian predators, as with the Jay which alighted near an Eyed Hawk moth (*Smerinthus Ocellatus*), saw the eye-pattern, was startled and flew away instead of eating the moth. Corvids are not averse to eating the eyes of carrion. Shepherds accuse Ravens and Crows of pecking out the eyes of dead lambs, and in Australia the first instinct of Ravens and Crows upon finding a dead lamb is to peck out the eyes and eat them.

Crows' sense of hearing is not as efficient as that of man, nor as that of some song-birds which have a hearing range of 20–20,000 cycles per second. For most birds the range is narrower and the Crow is limited to a range of 200–8,000 cycles which would not enable it to hear the lowest note on the piano which registers 27 cycles, although it could hear the 4,000 cycles of a piano's top note.

The loud, harsh calls of the *Corvidae* make them very conspicuous. As with all *Passeres* the syringial muscles of the syrinx are inserted on the ends of the bronchial semi-rings. The lower end of the wind-pipe has the last four or five rings welded to form a dice-shaped box and bronchial rings one and two are closely attached to this. The third ring forms an arched bar supporting a delicate sheet of membrane stretched between rings one and two on the one hand and four on the other. The bronchial rings' free ends support a tympanic membrane which plays an important part in voice production. A bony bar joins the bronchi with the wind-pipe and its free edge cuts across the bottom of the dice-shaped box of the wind-pipe, its vibrations acting like the 'free reed' of an organ pipe. Muscular lips extending from the inner surface of the bronchial semi-rings III narrow the aperture on either side of the 'reed' during the production of the voice, and thus complete the mechanism of voice production. Crows and more versatile song-birds have a highly developed syrinx, but in the case of the Crows it helps them greatly in mimicry and other calls.

Crows have their own communication calls which have recognisable meanings. Tame Corvids can be taught to say words. Konrad Lorenz found 'Roah', his pet Raven, was his only bird that could learn to use a human word in the correct context. Roah had an aversion to certain places and when his master walked in them he would bear down on him and fly over his head, then fly up, looking back to see whether Lorenz was following, and at the same time calling 'Roah!' with human intonation, obviously sensing that was his master's call note.

Distribution of the *Corvidae* is world-wide with the exception of southern South America, the Lesser Antilles, the Polynesian Islands, Madagascar and New Zealand, although the remains of what is thought to be an extinct Crow, *Paleocorax*, have been found in New Zealand, which may indicate that the *Corvus* group reached Australasia at a fairly remote period.

The family is thought to have originated in the now northern temperate and subtropical parts of the Old World and spread to its present range. The groups to which it is most closely related, viz Birds of Paradise, Orioles, etc occur only in the Old World and chiefly in the Tropics. Corvine fossils have been found in Miocene deposits in Europe dating back to 20–25 million years ago. The earliest known fossil, *Miocorax laiteti*, is from the Middle Miocene of Europe. The *Corvidae* fossils of North America date back to the early Pleistocene period. South America has a poor representation of *Corvidae*; Jays, the only species found there, have ventured no further south than 36°S latitude, which indicates that the *Corvidae* are of Old World origin. They are now well represented in the Palaearctic and Oriental regions, and the fact that the closely related groups, ie Birds of Paradise, are primarily tropical, inhabiting New Guinea and northeastern Australia, suggests that the *Corvidae* evolved in the Oriental region. Many forms occur in the Himalayan and Chinese mountains from which they could have evolved and spread north and south.

The Jays, the most primitive subdivision of the *Corvidae*, occur chiefly in the temperate north, and mountainous parts of the northern hemisphere; the genus *Garrulus*, the Common Jay, extends south well into the Oriental region, while the Szechwan Grey Jay, *Perisoreus internigrans*, is a native of the Szechwan mountains in China. Heinroth states that *Garrulus* has the least developed Corvid character of all the genera and that its behaviour resembles to some extent the bush-dwelling song-birds.

The Old World Jays have almost been eliminated by Magpies, Nutcrackers and Crows, which are all only transformed Jays. In the New World the Jays have less competition and they have undergone considerable adaptive radiation without losing their group characteristics. Jays entered the Americas at a time when tropical faunas extended further north than now and they were able to cross from Asia to North America via the Bering Land bridge. American Jays are predominantly tropical or subtropical, only six out of the thirty-two species ranging as far north as the United States.

Magpies are an Old World genus and are essentially large, heavy-billed, long-tailed, short-winged, often brightly coloured Jays. Their weak flight confines them to forest and bushy country. The New World besides having the Common Magpie, *Pica*, has Magpie-like types, ie Brown Jays, *Psilorhinus*, and Magpie Jays, *Calocitta*, whose relationship is so close that it makes a generic division difficult; whereas in the Old World, two natural groups, Jays and

2 The Common Jay *Garrulus glandarius* parent birds seen at the nest with young (*Eric Hosking*)

Magpies, are recognisable with a moderate gap between them. The Treepies, *Dendrocitta* and *Crypsirina* of south-east Asia, are nearest to *Pica*; they have about the same proportions, *Crypsirina* being smaller.

The Old World has produced a peculiar genus in the high, barren plateau of central Asia in the shape of the Ground Chough, *Podoces*. This small bird's terrestrial and cursorial habits provide a greater gap between it and *Garrulus* than the gap between the Jays and Magpies.

The Nutcracker, *Nucifraga*, is also a transformed Old World Jay and retains the white rump of *Garrulus*. *Podoces* and *Nucifraga* may have had a period of common ancestry after they diverted from Jays. *Nucifraga* reached America long enough ago for the American species, Clark's Nutcracker, *Nucifraga columbiana*, to become distinct from the Palaearctic species.

The Chough, *Pyrrhcorax*, in common with the Nutcracker, occurs in the Palaearctic region and inhabits mountainous country. The Chough, probably a further specialisation of the *Nucifraga* stock, is very specialised with a decurved bill, very long wings, booted tarsi and black plumage. The Crows are considered to be descendants of the Chough. They are similar in proportions, colour and, to some extent, habits. All *Corvidae* skeletons are much alike except in purely adaptive details. The skeletons of the Crows and the Magpies are almost identical except in size.

The Crow genus, *Corvus*, is the most successful and widespread of the *Corvidae*. It consists of the Raven, Carrion and Hooded Crows, Rook and Jackdaw. The Palaearctic, Oriental and Ethiopian regions contain many distinct species of *Corvus*. Crows with their superior colonising ability have colonised America and Australia at a more recent time, and have reached the Philippines, Polynesia and the West Indies, and other areas not otherwise occupied by *Corvidae*.

The genus *Corvus* is thought to have reached North America during the Caenozoic era, sometime before the Pleistocene period (1,000,000 BC), although the first fossils to be found date back only to the early Pleistocene period. There are similarities between the Eurasian *Corvus corone* and the American Common Crow, *C. brachyrhyncos*, and also among the Ravens of the Palaearctic and the Nearctic regions. Once the Asian Crow reached America it naturally flew southwards to warmer climes, but, not being a forest type, it did not penetrate the dense, unbroken forests of Central and South America. Strength is given to the opinion that Jays reached the New World before Crows by the fact that Jays have established themselves in South America in spite of Crows' superiority in flight and adaptability. Crows have established themselves in more island regions as far apart as the West Indies, Australia, the Philippines and small Pacific islands.

The earliest reference to Crows in Australia was made by Joseph Banks in 1770, when he mentioned they were shy. There are five species of *Corvus* recognised in Australia, consisting of two Crows and three Ravens. It is thought any invasion of this species into Australia was made from the north when a land bridge existed between New Guinea and Australia. Scientists say such land bridges existed prior to the Great Ice Age. In the late Pleistocene glaciation an ice cap covered the Kosciusko and Tasmanian highlands and sometime later the ice melted, then Tasmania became separated from Australia. Only the Tasmanian Raven, *C. tasmanicus*, is found in that island, which emphasises the relatively recent speciation that is progressively taking place in Australia.

Crow fossils of the Lower Pleistocene period (5–600,000 BC) have been found in the forest bed of Norfolk, Britain, where they shared the warm 'Cromerian' interglacial time with Mastodons, forest elephants and the first Mammoth. Animals of our time were also present and they remained during the Mindel glacial period (400,000 BC), when the Raven nested on crags and forest trees.

The Chough, *Pyrrhocorax*, is among the great number of birds' fossils found in Devon, Derbyshire and Yorkshire from the milder Eemian Interglacial period, (130,000 BC), through to the middle of the Upper Pleistocene when the species nested on the cliffs of South Wales. The Chough was much more widespread in Britain in the past than it is today. Gilbert White, the eighteenth-century naturalist, saw it on the Sussex cliffs and it was so common in Cornwall that it was called the Cornish Chough, but it no longer breeds there. The cold winters from 1820–80 in Britain synchronised with the Chough's disappearance. These birds would appear to suffer more mortality than other *Corvidae* in Britain as a result of inclement weather, probably due to their rather specialised feeding.

During the latter part of the Upper Pleistocene the avian fauna in Britain was very similar to that of today, but, in common with Ireland, it had a northern content. The Magpie co-existed with the Woodpecker in County Clare but it later became extinct in Ireland, to recolonise in Wexford and most of the country in, and after, the 1670s. In England it has increased since World War II.

Jay fossils are not mentioned until the Middle Stone Age when these birds were caught together with waterfowl by the Maglemosian hunters in Yorkshire, England. Their bones being smaller and more fragile than most of the *Corvidae*, their earlier fossils probably perished. Today the British Jay is a slightly more common bird since World War II and ventures into more public places.

Around 2,000 BC Ravens kept company with other Neolithic birds, and Crows were still widespread in Britain into the late Iron Age (200–100 BC). Their fossils have been found along with the Dalmation Pelican, waterfowl, and the Great Bustard, at the great lake village of Glastonbury in the Vale of Avalon, Somerset.

Ravens were tamed and taught to speak when the Romans

invaded Britain in 55 BC. Raven fossils of this period have been found in at least six settlements together with other species of the Crow family.

The Raven is the most widespread of the *Corvidae* family and occurs in practically all the arctic and temperate northern hemisphere. It is found throughout Europe, North Africa, Asia (except the southern and southeastern parts), Greenland, North America as far south as Nicaragua, and Australia. It has not fared well in populated areas because of its dislike of civilisation and carnivorous nature.

Up to 1,800 Ravens remained widespread as tree-nesters in lowland Britain, but during the Victorian era when game preservation reached its peak, it disappeared from its inland haunts in southern and eastern England and remained mainly as a rock-nester in the west and in Scotland. In Ireland it survived only on the rocky coasts and mountains. Its numbers have been severely reduced by man in many parts of its range and it has been eliminated altogether in some places. In western Britain it has made a slow recovery but the rock-nesting Ravens are fully established and over-populated there, forcing the new arrivals to return to tree-nesting.

Most Crows adapt well to man and thrive in civilised countries, but man is against them almost everywhere. In Britain, ever since 1457 when James II of Scotland passed an Act that ' "crawwys" should be destroyed and prevented from nesting in orchards and kirkyards' they have been considered pests, together with Rooks. In the sixteenth century the Statutes of Henry VIII and Elizabeth I made Crows vermin. In one English parish bounties were paid on 4,470 Crows' heads at a farthing a head. It would not be possible to kill 1,000 Crows in one parish so it must be assumed Crows and Rooks were regarded as Crows in the sixteenth century. In the early nineteenth century a lawsuit specifically mentioned Rooks. A case before the Probate Court concerned a man who was always changing his will and died leaving several claimants to his estate. He encouraged Rooks to his small rookery by having sticks scattered beneath the trees for their nests. This was given in evidence to prove he was of unsound mind.

In America the family *Corvidae* is now protected under the Migratory Bird Treaty with Mexico which was amended in 1972.

Under its terms the various species covered are protected during their breeding seasons except when causing damage or about to cause damage to crops or wildlife. Mexicans insisted on the inclusion of *Corvidae* as they have a number of species of Jays and Birds of Paradise, not found in the United States, that have become endangered. In the United States Crows and Magpies are in particular considered to be pests to crops. A decade or so ago roost blasting was rampant in the prairie states. But in spite of heavy persecution the Crows and Magpies continue to thrive. In Australia many methods to control Corvids from killing lambs have been devised but the most effective has been poisoning. This method is so simple there is great danger it may be used to the detriment of hygiene; the reduction of carrion-eating Corvids would lead to outbreaks of fly-strike and cause more losses of lambs than Raven predation.

In Britain today only the Raven (except in Argyll and Skye), and the Chough, are protected under the Protection of Birds Act 1954–1967. The rest of the British *Corvidae* come under the Second Schedule of the Acts and may be killed or taken by authorised persons. The Ministry of Agriculture used to organise wholesale shoots against Carrion Crows, the government subsidising cartridges used for this purpose. The shoots were referred to as 'Corbie-shoots'; 'Corbie' is the Scottish name for Raven but it has been used to describe the Carrion Crow of late. In the present days of inflation the cartridges are too expensive for these operations and the shoots have been discontinued.

The *Corvidae's* large consumption of grain in the British Isles brands them as pests, but to off-set this offence they eat corn-eating pests and weed seeds from arable land.

Ravens and Crows kill mammals such as rabbits, rodents and vermin. They dispense with carrion and prevent sheep and deer becoming infested with blowflies. Jackdaws perch on the back of sheep and delouse them. Magpies eat the eggs of game and domestic fowls for which they are ruthlessly shot by gamekeepers and farmers. Jays are also shot for this offence, although they eat mainly acorns. They are regarded as game birds over much of their range and shot for sport and food. They are fairly abundant in their favoured localities but, in common with all Palaearctic Jays, are overwhelmed by Magpies and Crows.

Indirect influences account for many casualties among the *Corvidae*. Damaging influences are to be found in indiscriminate felling of trees and destruction of woodland; development of natural land for building; and the change over of farmland from cereals to other crops. Such improved methods of harvesting as the combine harvester leave less loose grain on the fields. Coast-lines of cliffs and crags have been covered for miles with concrete thus depriving the Raven, Chough and Crow of their natural habitats.

Despite some men's aversion to the Crow family, it would be a sorry day if these handsome, adaptable and lovable knaves disappeared from the country. Each species of *Corvidae* has its own uses and not until man understands their intelligent, united life pattern and their place in the natural selection of the world's flora and fauna will they be fully appreciated.

Chapter 2

JAYS

Jays make up a very varied subfamily of 38 species, 32 of which are found in America, all but 4 of these being South American. The specialised feeding habits and nesting behaviour of some Jays suggest that their group is more ancient than the Crows and that they are more primitive than the rest of the *Corvidae*. They are more diverse and colourful than the Crows. They are found in all continents except Australia, and probably had a wider distribution that is slowly breaking up.

These bright-plumaged birds are smaller than Crows and generally have a relatively longer tail and shorter, more rounded wings. Many of them are crested. Their strong bills, which enable them to crack and hammer nuts, are about three-quarters the length of the head, not as long as the Crows'. Most Jays are arboreal and shy; they are usually seen in threes and fours but are not exactly gregarious. They hop instead of walk.

It is impossible to generalise about Jays as they have wide differences in behaviour and structure. Three genera have deviated somewhat from type and are placed first among the group. The Shrike Jay, *Platylophus galericulatus*, is found in the Tropics, its range extending from the Malay Peninsula to Sumatra, Java and Borneo. It was thought to have belonged to the shrikes (Laniidae), but its nesting and general habits are Jay-like. This genus has four species and may be primitive. The plumage of the Shrike Jay is black or brown with white crescents on the sides of the neck. A few of the head feathers are greatly elongated and broadened near their tips, forming a large crest. It has sparse and short nasal bristles but the rictal bristles (ie those fringing the gape) are exceptionally long. The bill is similar to that of *Garrulus*, the true Jays, but it is more hooked at the tip.

The White-winged Jay, *Platysmurus leucopterus*, in many ways connects the typical Magpies with the typical Jays. The genus contains two species which are to be found in Malay and Sumatra and the Borneo lowlands, occasionally up to 3,500 feet (1066.8m). This black Jay has white streaks on the wings. Its frontal erect crest has bristly and elongated feathers, and the long, broad head feathers form another short, full crest. The rictal bristles are long and the black bill is well curved with numerous short, stiff bristles covering the nostrils. The legs, feet and claws are black and the iris has a crimson hue. The well-graduated tail is not very long and less than the length of the leg. This Jay keeps to the forest in parties of 4–6. It is very restless, forever on the move high in the trees, uttering deep, rolling, metallic notes and living on fruit and insects. Breeding takes place in March and April. The rough, heavy, cup-shaped nest with a shallow interior is built of twigs and roots in bushes, small trees or palms, some 6–8ft (1.8–2.4m) from the ground, and 2–3 eggs (33.5 × 23.1mm) are laid. They are greyish or pale green on a ground colour with reddish-brown blotches, similar to the eggs of the Green Magpie of the same locality. The young are fed by both parents whom they resemble when they are fledged.

The Pinon Jay, *Gymnorhinus cyanocephalus*, is found on the drier slopes of the foothills on the mountains of the southwestern United States and northern Lower California among the pinon pines and junipers. This genus is an evolutionary offshot of the Crow family with no close relatives. Its form resembles the Crow with its short, squarish tail and strong, crow-like flight, but it is similar to the Jays in colour and habits. Its greyish-blue plumage is loose and fluffy; it has a pale rump and deeper colour on the crown region, with the sides of the head an almost azure-blue. From the chin to the centre of the chest the plumage is broadly streaked with greyish-white. This species has no crest or nasal bristles. The narrowing of the front of the skull and the long bill resemble the Nutcracker. The Jay feeds on the sweet nuts of the pinon cones and juniper berries; it also eats seeds of other pines, red cedar fruits and insects. In common with Crows it cannot resist birds' eggs and nestlings.

Like the White-winged Jay, the Pinon Jay is restless and wanders about erratically. It travels in flocks and is very sociable. This sociability extends into the breeding period which starts in February and continues into June. The birds fly to the Great Basin and build their nests in colonies with sometimes three nests in one tree. Occasionally they nest alone. The deep bulky nests, usually built from 2 to 18 feet (0.6096–5.5m) from the ground, are made of sticks and lined with weeds, dry grass and broken-up bark fibres which form a felt-like lining. The female, smaller and duller than the male, lays 3–5 bluish-white eggs, thickly and evenly spotted, which she incubates for 16 days. Both parents feed the nestlings on pinon seeds and insects. After three weeks the fledglings are duller versions of their parents and soon leave the nest to gather in flocks.

Pinon Jays make a wide variety of calls which range from a shrill, querulous 'peeh peeh' like the Arizona Jay to a 'chaar' as harsh as the Clarke's Nutcracker; sometimes they gabble like a Magpie.

The thirty-two species of American Jays are closely related, and belong to nine genera. Although primarily tropical, they are distinctly a North American group as South America has no genera peculiar to it alone; but it has two North American genera: *Cyanocitta* and *Cyanocorax*.

The genus *Cyanocitta* has two species: *Cyanocitta cristata*, the Blue Jay, with four subspecies; and *Cyanocitta stelleri*, Steller's Jay, with fourteen subspecies.

The startlingly blue wings and tail of the Blue Jay are a common sight in woodlands, parks, gardens and open forests of southern Canada and central and eastern United States, as far south as Florida. Except in the nesting season the loud calls of 'jay, jay', and whistled 'too-wheedle, too-wheedle' are familiar sounds, together with chattering and conversational notes. The bird also has a soft, pleasing, but infrequently heard song. It is a good mimic and can scream like a Red-shouldered Hawk. An elderly gentleman who raised Blue Jays found his best pupil could imitate the squeaking of his chair, the children knocking to come in and the family whistle for the dog.

Because of its intelligence and aggression this Jay has adapted itself to cities. It is a tyrannical visitor to the bird feeders in suburban gardens, routing the sparrows and starlings in order to devour the seeds and nuts. But I have seen it eating amicably with the Cardinal (*Cardinalis cardinalis*). While it was intent upon eating I noticed the purplish blue of its back, nape and crown, the latter with a pointed crest; and the black collar which extends from the back of the crest to the centre of the greyish breast, and its blue, greyish-white face. It has black stripes on its wings and tail, reminiscent of the Eurasian *Garrulus*.

The slightly smaller female incubates 4–6 pinkish-buff to greenish-blue eggs with brown or olive spots, for 17–18 days, in a nest made of sticks, moss and wool and lined with grass and feathers, 8in wide and 4in deep (20.3cm and 10.1cm). When the fledglings are three weeks old they are almost indistinguishable from their parents except for their shorter tails. They stay close to their parents upon leaving the nest and accompany them, chattering excitedly, in flight. To illustrate the young Jay's attachment to its elders is the story of a young Jay who fed an old grizzled Jay whose lower mandible was broken off near the base thus preventing him from picking up food. The young Jay flew to the old bird holding some food in its beak; the old Jay turned eagerly, lifted its crested head and accepted the food the obliging young one thrust down its throat.

At the end of the breeding season the Jays flock in hundreds, their raucous calls signalling their whereabouts. They migrate about one hundred miles south; many arrive on the south-west of Lake Erie in mid-October. This flocking and migrating coincides with the ripening of the nuts, berries and fruit upon which they gorge themselves. They are practically omnivorous; their staple diet is nuts, grains and fruit but they eat any small creatures, even bats, during the day time. They dispute over food, hold acorns in their feet and crack them with their black bills. They also bury acorns and help to vegetate the forests like *Garrulus* and share their Eurasian relatives habit of 'anting'.

Steller's Jay, *Cyanocitta stelleri*, is slightly larger than most North American Jays and is the only western species with a long, triangular crest. It is much darker than the Blue Jay; its head, neck, upper back, rump and under parts are deep blue and it has black bars on the wings and tail similar to those of the Blue Jay. One of the thirteen subspecies, *C.s. annectens*, is much paler and often has a whitish area over the eyes. It occupies the interior part of the Steller's Jay's habitat which extends from southern Alaska south-westwards as far as southwest Texas.

This subspecies betrays its whereabouts with harsh cries of

'Shack-shack-shack', and 'wek-wek-wek', as it gathers in small groups in conifer and mixed wood forests; it also ventures into orchards and gardens. It has other calls and a warbled song during mating and nesting in conifers such as Douglas firs. Apart from the nest being frequently plastered with mud it is very like that of the Blue Jay. The 3–5 greenish eggs are spotted brown or olive. The adults feed their young entirely on tent caterpillars thus ridding the continent to some degree of a pest that has reached disturbing proportions in some states. The young Jays have a dark plumage with blue wings and tail.

The size of this Jay's crest differs in various parts of the continent. The larger crest is more common in open country; it is raised in aggressive encounters and the larger crested birds dominate the social hierachies of their selected area, taking precedence over the other birds at feeding time.

The genus *Aphelocoma* belongs to the inornate line of Jay evolution which is thought to have taken place in the New World. The genus lacks the crest, black wing and tail bars of the Blue Jay, and the wings are shorter.

There are three main species of *Aphelocoma*: *Aphelocoma coerulescens*, the Scrub Jay; *Aphelocoma ultramarina*, the Mexican Jay, and *Aphelocoma unicolor*, the Central American or Unicoloured Jay.

The Scrub Jay has fifteen subspecies, one of which, *Aphelocoma coerulescens coerulescens*, the Florida Jay, is the smallest of the species and occurs only in Florida. It has the peculiarity of being one of the few birds with a range restricted to a smallish area. The other members of the genus are spread over the south-west United States and Mexico. This unusual gap in distribution suggests that this genus had a wider population in North America and is now breaking up.

The crown, rump, wings and tail of this genus are azure-blue, with the greyish white throat gradually shading into light grey on the chest. The blue feathers on the lower chest form a conspicuous semi-circular collar. The back is brown and the smoke-grey under parts fade to white on the anal region. The male is larger and has slightly brighter plumage than the female.

In Florida the Jay lives in brush country of scrub vegetation with small trees such as the wax myrtle on which to perch, nest, preen and feed. It concentrates along the edge of man-made clearings and perches on chimney-pots of summer cottages in the sand dunes. During a dry summer it is attracted to the sprinklers from which it catches drops of water.

Territorial aggression is strong in this Jay. It pairs territorially, and mated pairs screech and chase intruding immatures into the flocks in autumn. Its character varies according to habitat. The Florida Jay is interfering, inquisitive and fearless, and tame enough to perch on a man's hand. The Californian subspecies are also tame. Woodhouse's Jay, *Aphelocoma c. woodhousei*, of the Great Basin, is more furtive, while the subspecies in Arizona, *Aphelocoma c. nevadae*, is sly and suspicious.

This genus hops instead of walks and flies less strongly than the Blue Jay, although it can reach a height of some forty feet. It shares the Jay's eating habits. In spring it spends much of its time in the myrtle bushes deftly snatching insects before or during flight.

When scolding or excited the Florida Jay utters a harsh, grating call: 'kǎ', which sometimes almost rises to a screech. It also has a rattle call; 'krǎ', repeated rapidly during the frequent squabbles among its species. It expresses alarm by bobbing its head and flitting its tail while calling. It warbles in a 'whisper song' of sweet low-toned calls. The female threatens other Jays by slightly spreading and pressing her tail on the ground and thrusting her head upwards, accompanied by a hiccuping sound. She uses the alarm call when strangers approach.

Mating and breeding are in April. In courtship the female makes

3 Jays of the genera *Aphelocoma* and *Cyanolyca* (facing page)
a *Aphelocoma coerulescens coerulescens* of Florida
b *Cyanolyca cucullata mitrata* of Mexico
c *Cyanolyca viridicyana turcosa* of the temperate zone of the Andes, Colombia and Ecuador
d *Cyanolyca pulchra* of the upper tropical and subtropical zones of the Western Andes' western slope
e The Mexican Jay *Aphelocoma ultramarina* in 'rocking' posture during mobbing with *weet* calls
f The Mexican Jay *Aphelocoma ultramarina* in singing posture
g *Cyanolyca argentigula argentigula* of southern Costa Rica and western Panama
h *Cyanolyca mirabilis* of southwestern Mexico

a

b

c

d

e

f

g

h

begging notes as the male feeds her either on the ground or on the nest. Sometimes the male helps build the compact, flat nest in low scrub, but the female does most of the work. The pair desert a nest if it is interfered with. After hatching the 3–5 bluish or greenish eggs, spotted brown or black, the female broods and shields the nestlings from the sun until they grow older when she helps the male feed them. Non-breeders are occasionally allowed to help with the feeding, but the Scrub Jay, not being a colonial nester, is aggressive to other mature Jays. After twenty days the young leave the nest and join the autumn flocks.

Another subspecies, *Aphelocoma coerulescens californica*, the Californian Jay, resembles the Eurasian Jays, *Garrulus*, with its brownish-mouse-grey upper parts and white under parts. Around the back of the neck and the crown are cobalt blue with the sides of the head dark blue.

When nesting in the scrub and oak bushes in the dense chaparral of central California this Jay is very quiet, communicating in whispers and flitting silently from tree to tree.

Acorns are the Jay's main food but it is omnivorous and eats quantities of fruit, steals eggs, kills young chicks with its beak and flies after insects, catching them in the air.

One of the crestless Jays of the Rocky Mountains is *Aphelocoma c. woodhousei*, Woodhouse's Jay. Its plumage is a dull azure blue above with a brighter blue crown margined laterally with a narrow white streak. The under parts are greyish. It has a strident scream: 'Wry', and a variety of chuckles and gurgles when in the flock. In courtship the male's voice is as soft and insinuating as a dove's.

Over long distances this Jay has a steady flight, but during nesting it flaps and sails around the scrub oaks and low trees. In winter it gathers in noisy, restless flocks.

A mountain species of east-central Mexico, *Aphelocoma c. cyanotis*, the Blue-eared Jay, has the blue face of Woodhouse's Jay, but the back is more fequently tinged with blue and the rump is white. It also has a longer wing.

The behaviour of *Aphelocoma ultramarina arizona*, the Mexican Jay, differs greatly from that of the Scrub Jay. It is a communal nester and displays a social organisation shared by no other American bird north of the Tropics. Groups of 8–20 birds occupy dry oak woodland and defend a specified area against other groups, using a continuous guttural chatter to keep the flock together.

This Jay of the upper Sonoran life-zone of Arizona and Mexico is the largest species of *Aphelocoma*, and by reason of its flocking habits it has a different character and presence from the Scrub Jay. It is less active; its plumage is duller; it is not motivated to be aggressive and it lacks the rattle alarm call of the Scrub and Texan Jays. It also matures much more slowly; the bill of the first year bird only darkens on the upper mandible, and the lower mandible does not become wholly black until the third year, unlike the above, whose bills turn black almost as soon as they are fledged. This Jay's moult takes longer than other races of *Aphelocoma*, usually beginning in June and taking 4–6 months to complete. First year birds of this species do not breed.

The pair-bond is weaker in this species. Although, like the other species, the Mexican Jay circles round his mate in courtship with his wings and tail slightly spread, when he holds her he holds his bill down and fluffs his belly feathers, giving him a hunch-back appearance. Also he does not feed his mate away from the nest and sometimes other Jays participate in feeding the female and the young. The female often has outside help with building the nest 10–25 feet (3–7.5m) up a live oak tree. The blue, unspotted eggs are peculiar to the species. The female has no nest-defence when incubating but only gapes and 'Aahs' at other Jays as she would at her mate's approach. The Jays steal each others' nest material and eggs, but should an alien predator interfere a flock of Jays will mob it. They enjoy badgering a hawk and keep up an incessant chorus of loud 'Weet' calls. They also mob foxes and will alight near a rattlesnake, shrilling in a ludicrous manner with head and body stiffly upright and the end of the tail dragging the ground.

When in flight the Jay makes a singular fluttering sound with its wings, and its song, 'Weet, weet, weet', is delivered with the bill extended and virtually closed, the head held forward, turning from side to side, and the breast feathers fluffed. Although it is naturally timid, it enjoys prying into everything and the Mexican peasants called it the *Grajo ayul* (Blue Crow). It scavenges and snatches food from the logging camps of the Sierra Madre and from the picnickers, where it becomes tame enough to take food from their fingers.

There are seven subspecies of the Mexican Jay. Couch's Jay, *Aphelocoma couchi*, from southwestern Texas lacks the communal organisation of Arizona birds. It pairs and nests singly and has a rattling, aggressive call, resembling the 'Keep clear' of the Blue and Steller's Jays. Usually two nests are found over 100 metres apart; they are similar in size and structure to that of the Arizona Jay. This Jay's plumage is more like the Scrub Jay, the blue areas on head, rump, wings and tail are brighter than the Mexican Jay's; the grey-brown on the back is a darker shade, and the throat a more contrasting white. The bill resembles the Scrub Jay's, turning black soon after the young leave the nest. The 4–6 eggs are heavily spotted with dark green unlike those of the other species.

There are five subspecies of *Aphelocoma unicolor*, which occur in the cloud forests stretching from Arizona to Central America. Besides being the most southerly species of *Aphelocoma*, it is the most unfamiliar. It has a longer wing than the Scrub Jay and resembles the Mexican Jay in proportions, development of bill colouring, and gathering in flocks. The greenish yellow bill darkens after the first year to a more striking colour contrast than the Mexican Jay; it also has a slightly more resonant song.

Although more primitive, the genus *Cyanolica* is more closely related to *Aphelocoma* than any other genus. Both genera have a facial mask bordered by a variable white line separating mask from crown, and they usually have bristly or velvety frontal feathers. Both genera lack the crest and white-tipped tail of the ornate Jays and have similar calls.

There are seven species of *Cyanolica* with various subspecies. The Azure-hooded Jay, *C. cucullata* and the Black-throated Jay, *C. pumilo* are both birds of the cloud forests of Mexico and mix with each other on the ground among ferns and sweet gum, feeding on such trees as sycamore. *C. pumilo* has a distinctive black facial mask bordered by a thin white streak.

The Dwarf Jay, *Cyanolica nana*, a slightly smaller Jay, has a black streak passing from bill to eye but no black on crown or neck. Its range is smaller, but similar to the White-throated Jay, *Cyanolica mirabilis*, a black-faced species with a white streak dividing the forehead and ear coverts from the black head and neck. These species inhabit fir forests about 9,000 feet (2,743m) up in the

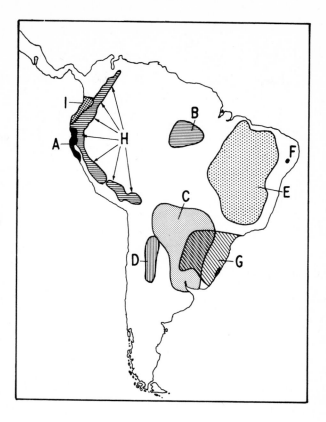

4 Distribution of South American Jays of the genera *Cyanocorax* and *Cyanolica*
  A *Cyanocorax mystacalis*
  B *Cyanocorax chrysops diesingii*
  C *Cyanocorax chrysops*
  D *Cyanocorax chrysops tucumanus*
  E *Cyanocorax chrysops cyanopogen*
  F *Cyanocorax chrysops interpositus*
  G *Cyanocorax caeruleus*
  H From north to south seven races of *Cyanolyca*:
  C. viridicyana meridana
  C. v. armillata
  C. v. quindiuna
  C. v. turcosa
  C. v. jolyaea
  C. v. cyanolaema
  C. v. viridicyana
  Cyanolyca pulchra

mountains of Sierra Madre del Sur: but *Cyanolica mirabilis* has the widest ecological range of the Mexican *Cyanolica* and penetrates the humid forests of Guerrero.

The closest living relative to *Cyanolica mirabilis* is thought to be the Silver-throated Jay, *Cyanolica argenticula* of similar plumage, but whose range is in Costa Rica and Panama.

A more ornate form is the Collared Jay, *Cyanolica viridicyana meridana*, which has a more complex repertoire. Its range is from the Venezuelan Andes down to Colombia, northern Ecuador, Peru and western Bolivia. The Beautiful Jay, *Cyanolica pulchra* with short tufted feathers on the forehead is confined to the upper tropical and subtropical zones of the Western Andes' western slope in southwest Colombia and west Ecuador. The vocal repertoire of *C.c. meridana* and the short tufted feathers on the foreheads of *C. pulchra* and *C. cucullata* suggest they are near living relatives of ancestral forms linking *Aphelocoma* and *Cyanolica*.

The nasal alarm call characteristic of *Aphelocoma* is found in *Cyanolica pumulo*, *cucullata* and *mirabilis*, the last having more than two basic calls. *Cyanolica nana* utters only one nasal 'perzheeup', similar to *Cyanolica mirabilis*, whose three variants of this call have a less highly pitched, fuller quality. In both species the call contacts the flock. The second call of *Cyanolica mirabilis* is uttered in couplets, and the third is a nasal alarm cry with a querulous, accentuated first syllable. *Cyanolica cucullata* has a loud, clear, whistling note.

The flock unity of *Cyanolica* resembles that of the Scrub Jay rather than the highly social Mexican and Central American *Aphelocoma* species. The birds usually group in pairs but sometimes 3–4 birds flock together.

5 Jays of Mexico and Central America (facing page)
  a *Calocitta formosa formosa* the Magpie Jay of Southern Mexico
  b *Calocitta formosa colliei* the Magpie Jay of Western Mexico
  c *Psilorhinus morio* the Brown Jay, of Mexico and Central America (*upper* head adult; *lower* head sub-adult; tails of white-tipped and plain-tipped species)
  d Genus *Cissilopha* the San Blas Jay (*upper* head *Cissilopha sanblasiana* adult; *middle* head adult of eastern races of *Cissilopha sanblasiana yucatanica* and sub-adult of *Cissilopha sanblasiana* and *melanocyanea*; *lower* head juvenile of *Cissilopha beechei left* tail normal for adults and sub-adults; *right* tail juvenile and first-year of *Cissilopha yucatanica*)

Cissilopha contains three species which are closely related and comprise a single super species. Their range is restricted from northern Mexico to Central America, and their closest relative is *Cyanolica pulchra* which has short, tufted feathers on the forehead, a heavy bill, and a mostly blackish head similar to *Cissilopha*. There is also a link with the ornate line, *Cyanocorax* in the voice and plumage of the Bushy-crested Jay, *Cissilopha melanocyanea*.

All the species of *Cissilopha* have an infusion of melanin which blackens their faces and obscures the facial pattern; in many this extends to the breast. In all three species there is a suggestion of a crest. The San Blas Jay, *Cissilopha sanblasiana*, has the most prominent crest of the species; it is composed of frontal nasal feathers up to 2cm long. In the Purplish-backed Jay, *Cissilopha beechei*, of Northwestern Mexico, the juvenile plumage crest appears as a prominent tufted growth above the eyes. This crest nearly disappears to form a prominent eyebrow in maturity and can be depressed and erected to express the bird's moods. *Cissilopha sanblasiana yucatanica* has a similar erect, stubbly crest, but in immaturity this species is snow-white except for the mantle, wings and white-tipped tail. It is spectacularly different from the other New World Jays in the colour of its plumage and in its yellow bill, which in other species darkens in maturity.

The voices of *Cissilopha* are rather complex. *Cissilopha beechei* and *Cissilopha melanocyanea* have a corvine cawing note. *Cissilopha beechei* also has a bell-like call and chatter similar to the ornate line, *Cyanocorax*, which it shares with *Cissilopha sanblasiana* and the subspecies *Cissilopha sanblasiana yucatanica*. *Cissilopha melancyania* has an additional variety of calls, some resonant and piping as in *Cyanocorax*.

The genus *Cyanocorax* is like *Cyanocitta* (Blue Jay), but in general more specialised. Its distribution is in Central and South America. Most species of *Cyanocorax* have a crest but this is extremely variable. The throat, breast, and usually large areas on the sides and top of the head, are black. The black breast is more sharply separated from the underside than in *Cyanocitta*. The lower breast and abdomen, and large areas of the tail, are characteristically white or yellow, but sometimes blue or purple. These Jays are larger than *Cyanocitta* and have stronger bodies, legs and feet. J. W. Hardy in his *Revision of the New World Jays* has divided *Cyanocorax* into four groups, with the Curl-crested Jay, *Cyanocorax cristatellus* as a monotype species.

In group I he places the Tufted Jay, *Cyanocorax dickeyi*, the White-tailed Jay, *Cyanocorax mysticalis*, the Plush-crested Jay, *Cyanocorax chrysops*, and the Black-crested Jay, *Cyanocorax affinis*. These species exhibit the oldest characteristics of body structure and colour pattern to be found in the ornate line. They have

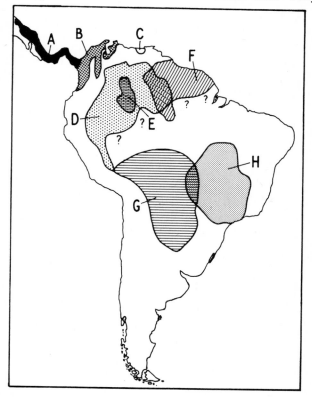

6 Distribution of Middle and South American Jays, *Cyanocorax*
  A *C. affinis zeledoni*
  B *C. affinis affinis*
  C *C. violaceus pallidus*
  D *C. violaceus violaceus*
  E *C. heilprini*
  F *C. cayanus*
  G *C. cyanomelas*
  H *C. cristatellus*

prominent tufted crests and a bold complex plumage pattern. Some of them have long tufts over the nostrils and forehead; their faces usually have a triangular white cheek patch and a spot or spots in the region of the eye. Many of them have white-tipped tails.

The vocal repertoire of this genus includes 4–12 distinctive call categories, with smaller sounds including the downward flexed 'Jay' or 'Jeer' call. The richer notes are a combination of sounds either bell-like or like a squeaking gate. *Cyanocorax mystacalis* of the open scrubland in southern Ecuador and northwestern Peru has a simpler repertoire than the other species.

Cyanocorax chrysops is unique in that the entire top of its head has bristly feathers like a bizarre crew cut and the whitish-blue feathers on the back of its head fit tightly below it. It has pale, peering eyes, and calls and bounces about excitedly, pumping its white-tipped tail.

Cyanocorax dickeyi, the Tufted Jay, has a rather isolated range in a small area on the Pacific slopes of Sierra Madre. It nests and feeds in the humid forests of the steep-sided river valleys and visits the mountain ridges for acorns in autumn. It tears apart epiphytic plants (plants other than parasitic ones growing on other plants) for seeds, fruits and insects. Sometimes it eats hanging upside down and sometimes catches green katydids in flight. *Cyanocorax dickeyi* is similar to *Cyanocorax mystacalis* of Ecuador although the two species are 3,000 miles apart. Both Jays have a white head, blue back and extensively white tail, that of *C. dickeyi* being longer; and the straight, stiff crest stands erect on the crown, dividing the top into tufts of stiff, narrow, elongated feathers. The small, territorial flocks of *C. dickeyi* number 4–16 birds and they remain in the same territory over a long period.

In late March a mated pair sing duets and members of the flock help them build their bulky nest, weaving a bowl measuring $5\frac{1}{2} \times 2\frac{1}{2}''$ (14 × 6.4cm) from vine tendrils and rootlets, and decorating the rim with fresh green leaves. When incubating the 3–5 bluish-white eggs the female gives the begging call and is fed by the male and other members of the flock, who also help to feed the nestlings later. The male guards the nest constantly and the female leaves it for food only in the morning. The pair remain together throughout the breeding season.

Cyanocorax affinis, in common with *Cyanocorax chrysops cyanopogon,* represents only a slight divergence from the characteristics of the primitive form incorporated in *Cyanocorax chrysops, mystacalis* and *dickeyi.* Its existence in the Caribbean lowlands suggest that an ancestral form was widespread in appropriate habitats between South America and the Sierra Madre, the present range of *C. dickeyi. Cyanocorax affinis* may be an indirect survivor of that ancestor for it is very like *Cyanocorax chrysops* in plumage, except that it lacks the short crest at the back of the skull. It has many calls some of which are shared by others of the ornate line.

Cyanocorax chrysops cyanopogon diverges from type. In addition to the frontal crest of the species it has long, broad feathers at the back of the head which form a short, full crest overhanging the white nape. These crests resemble the Malayasian White-winged Jay. *Cyanocorax chrysops tucumanas* of Gran Chaco's heavy rain forests, has a heavier bill and a blacker back than most of the *Cyanocorax* subspecies.

Cyanocorax cayanus, heilprini and *violaceus* of Group II show the most frequent evolutionary trend found in New World Jays, with characteristic shortening of the crest, infusion of melanin producing an obscure plumage pattern and simplification of the vocal range.

The Cayenne Jay, *Cyanocorax cayanus,* has a short, well-developed crest, prominent facial markings, brownish plumage and a white-tipped tail. It has at least six adult calls, and is closest to the ancestral type. The rarer Azure-naped Jay, *Cyanocorax heilprini,* is the intermediate species. It is darker than the Cayenne Jay with a shorter crest. As in *Cissilopha,* its facial pattern is obscured save for two dark purple bars on the cheek, and its purple tail has white tips. Its range in south Venezuela and northwest Brazil is surrounded by that of the Violaceous Jay, *Cyanocorax violaceus,* the last of the species thought to have evolved. It has an even stubbier crest, dull black and brown plumage, and a plain tail. It has one call-note.

Group III has one species, the Green Jay, *Cyanocorax yncas.* This old independent form is similar to the three previous species. It is slender with a tail relatively longer than the former group. The green tail has bright yellow under parts. The population of *yncas* in central and South America ranges from west–central Mexico and extreme southern Texas, south to the highlands of Guatemala and

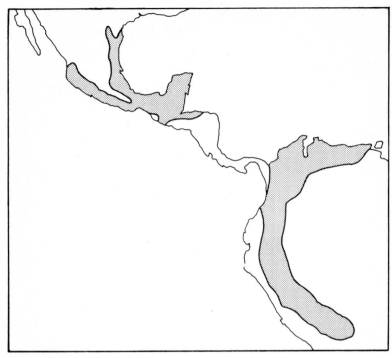

7 Distribution of *Cyanocorax yncas* in Middle and South America

northern Honduras; and from the subtropical zones of Colombia
and northern Venezuela south to eastern Ecuador, Peru, and
northern Bolivia. The population of *C. yncas yncas* in central South
America, including *Cyanocorax yncas*, is less greenish and yellow
and more bluish-green than the northern population and the
prominent frontal crest is larger. These species are thought to be
closest to the ancestral region of origin of the ornate line.

The Rio Grande Green Jay, *Cyanocorax yncas luxuosus*, has blue-
green upper parts and pale green under parts; the wings are short
and rounded and the head uncrested. It has nasal tufts and broad
cheek patches extending to the back of the lower eyelid, and its
crown, back of the head and nape are blue, although the forehead is
white. The chin, throat, chest and oral region are black. This
species is the only green-plumaged Jay. It previously inhabited the
woodland edges of plantations of the Rio Grande, but since they
were cleared for agriculture it has sought sanctuary in Santa Anna

Refuge near McAllen, Texas. Being a very inquisitive bird it ventures into the outskirts of towns and enters tents and houses for food. It has harsh, high-pitched calls, its common call being a long note followed by three short ones similar to the *Cyanocorax dickeyi* group.

In April and May the Rio Grande Jay nests in dense thickets of such trees as Mesquite, Retama, Brazil and Hackberry. It builds its flimsy nest on a platform of thorny twigs 8 × 8″ (20.4 × 20.4cm) across and 4″ (10.2cm) deep. The 3–5 eggs are blotched brown or grey. The young are fed on eggs and young of other birds besides seeds, insects and acorns. In winter these Jays favour ebony seeds and the Texas palmetto fruit. They are solitary nesters but flock outside the breeding season. Sometimes they have two broods a season. The young are duller and paler than their parents.

Group IV contains *Cyanocorax cyanomelas* and *C. caeruleus* of central and east South America. Both species are distinguished by their robust forms, black heads, short crests, and lack of white on the tails and abdomen. *Cyanocorax caeruleus* has longer frontal tufts. The obscure plumage pattern of *C. cyanomelas* is similar to *Psilorhinus*, the Brown Jay. *C. cyanomelas* is seen in parties with *Cyanocorax chrysops* and hops freely among picnickers. It has a loud crowlike 'carr-r-r' and 'chah chah' and *Cyanocorax caeruleus* has the same downward inflected note. Although *Cyanocorax cyanomelas* has the straight, direct flight of other Jays it has a peculiarity of progressing by slow beats of the wings, followed by half a dozen quicker strokes, ending in a long upward glide to its perch.

The monotype species, *Cyanocorax cristatellus*, is last of the group. It occurs in the Brazilian tableland to which no other species of the line has adapted. Its uniquely long wings adapt it to Brazil's short tree savannah. It has a long wing-to-tail ratio, a peculiarly curly crest, and the unique combination of a patterned body and tail plumage with a brown to black head, neck and upper breast. Its unique structural features show its early evolutionary separation and its long, independent development. Its one prominent call is a downward 'jeer' note.

There are four species of the specialised Brown Jay, *Psilorhinus* of Mexico and Central America, which deviates from the Jays in having a crop-like furcular (forked) sac within the collar-bone, which it uses to produce a snapping sound. Many other Jays can utter mechanical sounds apparently originating from the respiratory, instead of the vocal organs although no affinity of structure has yet been discovered. Probably some air sac mechanism with a similar function to the forked sac of the Brown Jay will come to light. Another extreme characteristic of the Brown Jay is its lack of blue coloration. It is brown with its paler colours whitish or yellowish. But its plumage is typical of the ornate line and the tail is white-tipped in some species.

In *Psilorhinus morio vociferus* of the northern Yucatan Peninsula, however, there is a largely random occurrence of white-tipped and brown-tailed birds. The former has been recorded as far north as Plan del Rio, Veracruz, and the brown-tailed south to Tabasco. The short crest and sparse nasal bristles of the Brown Jay, together with its obscure plumage pattern, and faint facial markings, ally it with *Cyanocorax cyanomelas* but its coarser structure resembles *Cissilopha*. Its highly social sense, and the helpers allowed at the nest, are characteristic of *Cyanocorax dickeyi* and *Cyanocorax chrysops*; so *Psilorhinus* would therefore seem to be an intermediate genus between *Cissilopha* and *Cyanocorax* with the furcular sac being its outstanding character.

The most bizarre member of the Crow family is the Magpie Jay of the genus *Calocitta*. It is thought to have replaced the earlier, more primitive Mexican species of *Cyanocorax*. The nominate sub-species *Calocitta formosa formosa* is a large bird, 42″ (106.7cm) long, with a 28″ (71cm) bright blue, graduated tail, which undulates with a wave-like motion when the bird flies, revealing extensively white-tipped side feathers. It has brilliant azure-blue plumage above, white below, a black throat and chest and a long crest bearing a similarity to the Blue Jay. But *Calocitta formosa collei* shows affinity with *Cyanocorax* with its black throat and chest, black face and ear-coverts framing blue cheeks, and its crown capped with a black, white-tipped crest consisting of very long, somewhat spatulate black feathers. The subspecies, *Calocitta f. azures* and *pompata* have white throats, thin black necklaces and white on the face. In all species the sexes are broadly similar but the females have more black on the crown, forehead and crest.

The Magpie Jay is a lowland species and seldom ventures above

4,oooft (1,219m). It is not migratory but occasionally visits the borders of southern Arizona, some distance from its habitat in the arid scrub-forests on the dry Pacific side of Central America from Mexico to Costa Rica. *Calocitta f. collei* extends its range to the state of Sonora, and *Calocitta f. pompata* has an isolated range on the Atlantic side of Guatemala.

This noisy, extroverted Jay mobs human intruders with verve and gusto during its very long nesting season, extending from November to July. Although it has a wide variety of harsh calls it utters a surprisingly sweet gurgling sound during the mating season. The birds build an untidy nest of twigs enclosing an oval cup of hair and fibres some 100ft (30.5m) up a tree. The female lays 3-4 grey eggs flecked with brown and is fed by her mate and favoured birds, while incubating. The parent birds are very aggressive to alien intruders and mob any boat-tailed grackle hovering nearby. This Jay is an omnivorous eater. It walks with an active, jaunty gait with its tail held high, and during the dry season it perches on cattle and relieves them of infestations of ticks.

The genus *Garrulus* contains the true Jays of which there are three species in Europe, Asia and northwest Africa. The specific features of this genus are the streaked crown, blue wing coverts and white rump. The allied forms differ in measurements and shades of colour. *Garrulus glandarius* is the only European species and has many subspecies.

The almost square, black tail and white rump of the British Jay, *Garrulus glandarius rufitergum*, are conspicuous when, at man's approach its rounded wings bear it away with a weak, laboured and rather jerky, flapping motion. In early spring this shy bird is recognisable high in the woodland's bare branches by its soft, full, pinkish-brown plumage, the blue and black barred alula, primary and outer coverts, and the white on the basal half of the first five secondaries. Closer inspection will reveal the streaked black and white erectile crown feathers; the broad, black, moustachial stripe; the white chin and throat, and the brown-black primaries. The fifth and sixth primaries are the largest, the first primary being half their length. The strong, black bill is about half the length of the bird's head; its rounded nostrils are bristly and its pale blue eyes surrounded by a narrow inner rim. It has pale brown feet and legs.

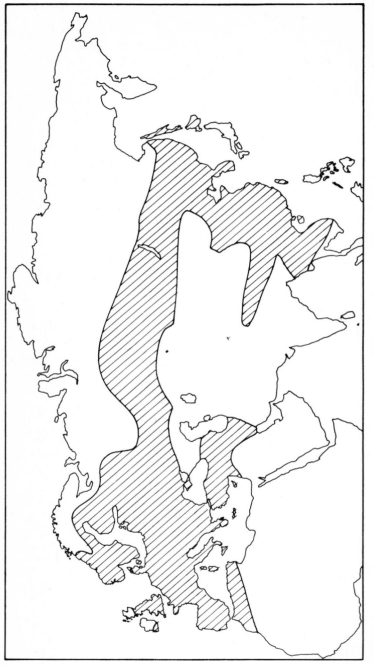

8 Distribution of the Common Jay *Garrulus glandarius*

When flying in open woods this Jay shows surprising dexterity, but it hops rather heavily both on branches and on the ground, jerking its tail up and down and from side to side. In alarm it emits a very harsh 'skraaak' and occasionally a raucous, far-carrying 'ro' or 'rah', rather like a heron's call. It also gives a loud, crow-like 'kobra-ra', and has a prolonged mewing note. It chuckles in flight.

Jays do not start breeding until they are three years old. They choose deciduous, mixed and coniferous woods, orchards and plantations in which to build their nest of only about 1ft (30cm) across with a fairly deep well. They never use old nests. The nest, made of twigs, and lined with grass, earth, fine roots and hair, is built 6.5–19.5ft (2–6m) above the ground, around the end of April when the trees are in leaf. Their courtship display consists of posturing and spreading wings and tail and the female begs the male to feed her. After the birds have mated the domination by the male is reversed. The female lays 5–7 glossy, oval eggs, 1.2 × 0.9″ (3 × 2.3cm), greyish or olive green with brown specks. She starts the 16 days' incubating, forcing the male to take his turn, and also to feed her.

Jay nestlings are naked when hatched. They are fed by both parents for twenty days, mainly on defoliating caterpillars which the parents store in their crops and regurgitate, pushing them down the nestlings' pale pink gapes which are flanged with whitish-pink. During this period the family remain quiet. When man approaches the Jay, unlike other Corvids, remains crouching in the nest with a partly open bill, but she flies off when he is within a few yards, silently, or with a few screeches which other Jays may accompany. Unless the parents desert the nest through human interference, there is only one brood. Jays do not desert when they have nestlings; they have a close social bond with their young and still feed them when they leave the nest after three weeks. Throughout the autumn the young remain with their parents who are never aggressive towards them, even when they have matured. The young are redder on the upper parts and breast than the adults and have smaller dark streaks on their forehead and crown. In a study of their breeding areas Jays were found to have lost on average 40–80% of their eggs or nestlings. Predators steal their eggs; besides human predators, Crows and squirrels are the chief predators. Both *Garrulus* and

Cyanocitta mob Crows and they appear to regard black Corvids as potential nest enemies.

The parents start to moult their primaries when they still have young in the nest, and their entire moult takes about 102 days.

Like the Blue Jay, the Common Jay with young will boldly attack hawks. They give the 'hawk alarm' at the sight of a sparrow hawk, looking about by moving their eyes but without moving their heads; then with a screech of alarm they puff out their body feathers in threat display with the head and neck feathers tightly down, while they jerk the head and body up and down, then screeching, they mob the hawk. They are great predator imitators. They can chatter like squirrels; make a popping noise, possibly imitating a gun, and give the 'kik-kik-kik' call of kestrels. They give an 'owl-screech' when attacking an owl and fly repeatedly at it. Jays react to other Jays' alarm calls, if they cannot see the danger; they also appear to recognise other birds' alarm calls. They mob man only if he gives them cause for alarm and they then 'mob' at a safe distance under dense cover.

In the spring 3–30 Jays will gather for what are thought to be 'Crow marriages'. The majority are mated birds but the unpaired birds pair by chasing each other excitedly along branches, with slowly flapping wings and 'tip-tip' chuckles occasionally inter-mingled with soft, melodious warbling. When they drop to earth they partially spread their wings and tail in the mating display. They also have noisy winter gatherings of up to a hundred birds which spring and dash about flirting their wings and tail and performing slow motion flights. When on the ground they always cross open country one at a time, the second bird waiting until the first is half-way across – a characteristic of birds with weakish flight that seek cover when alarmed. When in an aggressive posture the Jay stands sideways and ruffles and spreads his wings to appear large, but he points his bill up in his appeasement display to show he is ready to withdraw.

Acorns form the largest single item of most Jays' diet. They take them from the tree while still green, hold them in their inner toes while gripping the branch with the outer toes and hind toe; after hammering and levering the acorn they pierce it with their bill and remove the nut. A Jay can carry 1–3 acorns in its crop, and another

9 The Blue Jay (*Eric Hosking*)

in its bill, prior to burying them. Like some North American Jays, it takes the acorns some way from the parent tree, thus being a primary agent in regenerating the oak. In an oakwood thirty Jays were observed to bury some 200,000 acorns during the month of October and naturally they do not retrieve them all. If another bird watches a Jay burying or hiding food it will attack it before taking the food to a new hiding-place. It is perceptive enough to hide perishable food in crevices instead of burying it. It turns over loose leaves, opens crevices and small holes for insects, wedging holes open with its bill. Jays appear to have a sense of taste in their rejection of certain beetles such as the red and black soldier beetles (*Terephorus*). In the spring the Jay feeds on caterpillars but dislikes slugs and snails. It holds sticky objects in its bill and rubs them on the ground in order not to foul its plumage. It does not suck eggs but pecks at them and tries to swallow them. It has very keen eyesight and can detect the smallest insect 6ft (1.83m) away; it is a valuable destroyer of insects. Although fruits are taken and stored in their crops, Jays are the only Corvids in Britain that have little effect on agriculture.

Enjoyment, coupled with a sense of humour, features strongly in a Jay's life. In captivity Jays have been known to drop balls to see them bounce, and to hide coins. Great pleasure is obtained by 'anting'. The bird squats over ants, holding its body upright, with head slightly lowered and wings spread and thrust forward so that the primaries touch the ground. To contact the ants it drags its under parts along the ground, and gives convulsive shuddering spasms, continually bringing its tail forward and sitting on it. It runs its bill down the inner edge of its primaries and secondaries but it does not pick up the ants like most birds; in fact it dislikes the worker ants. Ants swarm up the tail, legs, wings and over the body; if one squirts formic acid in the bird's eyes the Jay closes them and hops away as though in pain. It takes a vigorous bath after 'anting' and then preens. Baths are enjoyed two or three times a day, but 'anting' is performed at intervals of two or three days.

Jays' gift of mimicry makes them amusing pets. A Jay named Jasper was kept in an aviary near a house. He had fifty sounds in his repertoire and related them to objects. He imitated the shrill call of guinea-pigs and their young master's voice. He copied the daughter's squeaking pail when she brought water to the aviaries. When the cat passed he 'meowed' and called 'Come here, Jason' to the dog, like the daughter. Although he only had a fleeting view of the rag-and-bone man he called 'Rag-bone' at him.

Jays move around in search of food more than Magpies. They are resident and locally abundant in England and Wales, and are said to have increased since World War II. Since a century ago they have decreased in Scotland but are gradually returning locally except in the north. They have been recorded in the Shetlands but are not on the Isle of Man. They are becoming less shy and moving closer to towns, visiting gardens for food and colonising in large suburban gardens and London parks.

The Irish Jay, *Garrulus glandarius hibernicus*, is darker and more rufous than the English species. It has increased and extended its range. It breeds all over Ireland except in the northwest. The European Jay, *Garrulus g. glandarius*, is similar to the British Jay but slightly larger, greyer and less shy. It is less pink on the back and has a rather paler breast and flanks. The streaking of its crown is variable. It occurs in the open conifer forests from Scandinavia to central and southern Russia, the Pyrenees, Switzerland, Austria and Hungary. In Romania and Bulgaria it intergrades with *G.g. cretorum* of Greece and Crete, and in Finland it intergrades with *G.g. sewerzowi* of northern Russia. *G.g. glandarius* migrates to England occasionally in winter, probably when the continental acorn crop is scarce.

There are only three subspecies of *Garrulus* in North Africa. *Garrulus glandarius whitakeri* of northern Morocco and western Algeria, *G.g. minor* of the Sahara Atlas mountains and the Middle and Grand Atlas mountains, and *G.g. cervicalis* of northern Tunisia and Algeria. The latter and *G.g. atricapillus* of Syria and southwest Persia both have black crowns. In the Middle East *G.g. atricapillus* intergrades with *G.g. krynicki* in northern Iran, and a secondary intergradation takes place with *G.g. krynicki* and *G.g. hyrcanus* in the region of Lenkoran, USSR on the Caspian Sea.

In various forms the subspecies *G.g. brandti* stretches across northern Russia, the Urals, eastwards through the Siberian taiga to northeastern Mongolia, northern and central Manchuria, Korea and Sakhalin, Hokkaido and the southern Kuriles Islands.

This species takes the form of *G.g. pallidifrons* in Hokkaido, where its harsh 'jaaarr' and clicking mewing noises call attention to its pale body and reddish crown in the forests, and to the small flocks in autumn that migrate to the towns. Another Japanese Jay, *Garrulus f. japonicus* is distinguished by a white crown streaked black, a bold moustachial stripe starting above the bill, and a pure white throat, abdomen and rump. Its wings are blacker than those of the European Jay and have a white patch, similar to *G. lanceolatus* of eastern Afghanistan. Other species in Japan are *G.g. tokugawee* of Sado Island on the northeast coast of Honshu; *G.g. hiugaensis* of the Izu peninsula, southern Kyushu and Kagoshima, and *G.g. orii* of Yakushima Island.

Further south in the East China Sea the specialised species, *Garrulus lidthi*, has an isolated range in the two small tropical islands of Amami Oshima and Taku-no-Shima in the northern Ryu-Kyus group of islands. It is a common resident in the dense tropical woods and pine forests near cultivated fields and villages. This Jay is larger than average, 15in (38cm). It has a black head with a metallic purplish crown, ear and nape; a dark blue chin and

throat streaked with white. Both upper and under parts are reddish-
chestnut, with the bluish-black wings having the alula, and sometimes the flight and long tail feathers, blue barred with black. Both
wings and tail have white on the feather tips. The bill is ivory white.
Although it is more magpie-like than *G. glandarius* it resembles this
species in flight, voice and habits, but it is most closely related to
G. lanceolatus of the western Himalayas. Besides acorns it takes
sweet potatoes from the cultivated areas and also eats small reptiles
and insects.

Another uncommon feature of Lidth's Jay is its substantial nest
in the hollow of trees about 3.3 × 13ft (1–4m) above ground and
usually with two entrances. It lays 3–4 greenish-blue eggs in March
and April. In the nonbreeding season 5–6 Jays flock together.

Japan has protected this Jay by law as a National Monument
since 1921 because of its interesting local distribution.

In India *Garrulus g. bispecularis*, the West Himalayan Red-
crowned Jay, is resident in the mixed forests of pine, oak, chestnut
and rhododendron in the Himalayas from West Pakistan, eastwards
to central Nepal where it intergrades with *Garrulus g. interstinctus*,
the Eastern Himalayan Jay. It has the pinkish brown plumage of
the European Jay with black barred bright blue wings and a black
tail, but its iris is pinkish red to dark brown with a plumbeous
orbicular skin.

This Jay has the Crow's characteristic habit of rallying the flock
and raising a hue and cry when one of its members is in distress;
it is also inquisitive, clannish, and both bold and wary like the Crow.
Noisy pairs or parties of four or five collect with harsh, penetrating
cries of 'shak'. They join larger flocks of twenty or more in autumn
and winter and associate with *G. lanceolatus* and the Himalayan
Treepie, from whom they copy raucous croaks and chuckles. They
also imitate the whistling squeals of the Blue Magpie and the cry of
such birds as the Hawk Eagle (*Spizaetus*) and the Indian Mynah
(*Acridotheres tristis*). They also have a rather low and varied song.

From March to June the Jays breed in loose groups of six or
more pairs with nests sometimes only twenty yards apart. Both
sexes build and incubate the 3–5 olive-green or olive-brown
speckled eggs and tend the young. In characteristic Jay fashion they
eat acorns and steal eggs and nestlings. They also hang around the

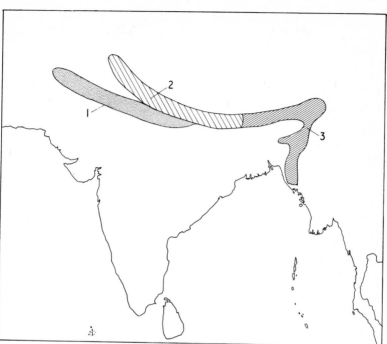

10 Distribution of Jays
  1 Black-throated Jay *Garrulus lanceolatus*
  2 West Himalayan Red-crowned Jay *Garrulus glandarius bispecularis*
  3 East Himalayan Red-crowned Jay *Garrulus glandarius interstinctus*

bungalows of hill stations for scraps and raid the hill orchards for fruit.

From eastern Nepal eastwards as far as the Assam Hills and southwards to East Pakistan and Manipur the East Himalayan Red-crowned Jay is resident and rather local in the wet temperate oak and conifer forests at an altitude of 6,562–9,843ft (2,000–3,000m) in summer, descending in winter to 4,921ft (1,500m) and extending into the semi-tropical zone. In appearance this Jay is similar to the West Himalayan species except that its tint is more rufous and its hazel brown iris is beaded with a magenta coloured ring.

The Black-throated Jay, *Garrulus lanceolatus*, is a distinct species and an intermediate between *Garrulus glandarius* and *Garrulus lidthi*. It has vinous grey plumage, a black tufted head, and a black chin, throat and foreneck boldly streaked with white. Its black wings and tail are barred with bright blue with a white patch on the front. It has steel-grey legs and bill, and its iris is either brown or vinous red. It is found in mixed oak and conifer forests in eastern Afghanistan, Northwest Frontier Province, and western Himalayas eastwards to Nepal. Both sexes build an untidy, deep nest in the top of oaks and saplings, and incubate the 3–5 stone or greenish white eggs, mottled with pale sepia brown. The young are fed by both parents and they have a dimmer plumage with a shorter crest. The noisy excitability of these Jays is well-known in hillside station gardens and the backyards of bungalows. They mob hawks and owls, swearing angrily and raising their crests and flicking their wings and tails.

The Jay population of Burma and Assam belongs to two types connected by intermediate forms. The Burmese Jay, *G.g. leucotis*, is a pure type. It is uniformly black on the centre and back of the crown, with a black crest; and white on forehead, ear coverts and throat. It has a broad moustachial streak. Its body is vinous brown, and its tail black. The primaries are partly grey; the greater wing coverts, the alula and the outer webs of most secondaries are bright blue, banded with black. It occurs 4,000–7,000ft (1,219–2,134m) up in pine and dry deciduous forests from Burma, south and east as far as the Mekong delta, South Vietnam. From March to the end of May pairs nest about twenty yards apart in communities of six or more pairs.

The other pure species is *G.g. persaturatus* of the Khasia Hills, Assam. It is the darkest and brownest of all the races and has no black streaks on the crown. Apart from its being shyer and less noisy its habits are similar to those of the European Jay. The natives call it '*Dao-flampu*'.

There are three intermediate types. Sharpe's Jay, *G.g. oatesi*, of the upper Chindwin and Chin Hills, is similar to the Burmese Jay but has the front of its crown and crest white, broadly streaked with black. In April it nests in the low tree and scrub jungle, 3,500–5,000ft (1,067–1,524m) up in the Chin Hills. The hen is noted for her constancy in remaining on her shallow nest when alarmed. Rippon's Jay, *G.g. haringtoni* is a large bird of the southern Chin Hills. It has faint brownish streaks on its crown, with buff forehead, ear coverts and throat. The other intermediate form is *G.g. azureit-*

inctus, which is more vinous than Rippon's Jay but paler than *G.g. persaturatus*.

The Yunnan Jay, *G.g. rufescens* of southwest China, adjoining Burma, is a subspecies of the Chinese species *G.g. sinensis*, and extends into the Shan States. It is rich vinous above, with a white throat, the sides of the head paler than the body and the forehead faintly streaked vinous.

The Old World genus *Perisoreus* is rather primitive. Its unspecialised appearance of greyish plumage and lack of a crest are thought to have been derived from its northerly habitat. Its range in Eurasia and North America is further north than the other Jays.

In Europe the Siberian Jay, *Perisoreus infaustus*, is found in thick northern coniferous and birch woods in central and northern Scandinavia and Russia. In Asia it extends from Siberia to Sakhalin, and south to northern Mongolia. This Jay has ten species. It is smaller than the Common Jay with an apportionately longer tail, more dense and fluffy plumage, and a shorter, black bill. Its wings, back, and under parts are mouse-grey but it is conspicuous in flight because of the fox-red on its rump, the outer feathers of its well-graduated tail and the patch on its wings. It has a dark brown head with a paler chin and throat.

These Jays live in pairs or small family parties. They show little fear of man and are confident and perky as they scavenge around camp sites. They eat conifer seeds, clinging to the tips of pine branches to reach the cones. They also eat berries, insects and, characteristically, eggs and nestlings. In winter they eat tree-growing lichen. In common with *Garrulus* they hide food. Although usually non-migratory they will fly to central Europe if food is scarce in their region. Their flight is similar to *Garrulus* but they are lighter on the wing. Regular wing beats alternate with short glides and their tail and wings are outspread. Their many calls can be delightfully musical but are sometimes the reverse. They also have a soft mewing note and indulge in mimicry.

Breeding takes place in the very cold months of April and early May. They insulate their nests by making the walls soft and thick and lining them with lichens, bark fibres, feathers, moss, hair and fur. The 3–5 greenish-grey eggs, blotched brown with underlying greyish-mauve markings, are incubated by the female. In common with Sharpe's Jay she sits tight on her nest at the approach of man and even trustfully allows herself to be lifted from it.

During the breeding season the parents are quiet and retiring. The young are adapted to the cold by being more thickly feathered than the Common Jay; they huddle together for warmth in an early frost and by fluffing out their thick plumage they can insulate themselves by trapping their own body heat.

The Szechwan Jay, *Perisoreus internigrans*, of the Szechwan mountains of western China, resembles *Garrulus lanceolatus* in its bill and black head which strengthens the belief that *Corvidae* originated in the mid-Asian mountains. This is a larger bird than the Siberian Jay and has a stronger bill and legs. The appearance of both *P. infaustus* and *P. internigrans* in Asia indicates that *Perisoreus* is an Old World genus which reached America later than the other Jays in the form of *Perisoreus canadensis*. This genus, the Grey Jays of North America, has nine species and is found from north–central Alaska across Canada to central New York, northern Vermont, New Hampshire and Maine, and south as far as northern California, central Arizona and northern New Mexico on the west of the continent.

The only relief to this Jay's grey, loose, fluffy plumage is its white forehead, face and chin, which stand out against its dark hind-neck, brown eye and dark, short bill. Its bill, the lack of a crest, longish tail and rounded wings resemble the Siberian Jay but it has occasional white on the larger wing coverts and its anal region. It boldly visits camps and picnic grounds for scraps, and cleverly stores acorns and meat in trees when the ground is covered with snow. It sticks food in the hiding-place with its own saliva, supplied by special enlarged saliva glands. Like other Jays, it is a mimic and has a varied repertoire. It can chatter, whistle: 'quee-oo', chuckle and give a soft, prolonged warbling song.

This Jay builds its nest near the tree-line in coniferous and mixed wood forests between 4–15ft (1.2–4.6m) above the ground. Its bulky nest is made of the usual materials with the addition of caterpillars' cocoons. The female incubates the 2–6 pearly grey eggs spotted olive buff for 16–18 days. The young's first plumage is slate colour later changing to blackish grey. In winter the Jays fly to lower altitudes within their breeding range.

Chapter 3

MAGPIES

The Old World Magpies can be recognised as a separate group from
the Jays, the gap between the two genera being narrowly bridged
by Lidth's Jay, which is similar to the showy Ceylon Blue Magpie.
Both birds have a rich chestnut back and breast and long tail
feathers of blue with black tips or bars.

The Old World genus *Pica*, the Common Magpie, reached
North America relatively recently and will probably adjust itself to
eastern North America, since the Old World species occur in both
semi-arid and humid regions. In the New World the intergradation
with other Magpies and Jays is so complete that it has been difficult
to form a generic division and all the species have been assigned
to Jays.

The Magpie was originally known as a 'Pie' on account of its pied
plumage; the feminine prefix 'Mag', meaning 'chatterer', was
added about the end of the sixteenth century.

The skeletons of *Pica* and *Corvus* are almost identical except in
size. They may be closely related but if the Crow evolved from the
Magpie it must have been rather early in Magpie history for the
birds to have acquired such different proportions. Magpies are
large, heavy-billed, long-tailed, short-winged, often brightly
coloured birds, their weak flight adapted to forested or bushy
country. There are four genera of Magpies: *Pica*, *Cyanopica*,
Cissa, *Urocissa*, and three genera of Treepies: *Crypsirina*, *Dendrocitta* and *Temnurus*. *Pica* and *Cyanopica* are primarily Palaearctic but
extend into the northern parts of the Oriental region. *Cissa* and
Urocissa together with the Treepies are Oriental in distribution.

The species, *Pica pica*, the common black-billed Magpie ranges
from Anadyrland in north-east Asia, to central Asia as far south as
southern China, Tibet, and westward across Europe and north-west
Africa. In North America it extends from central coastal Alaska

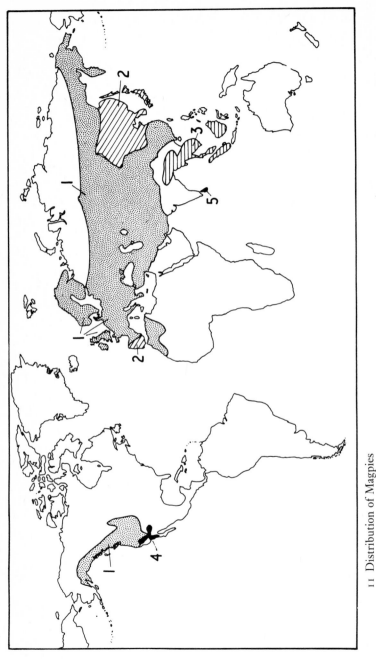

11 Distribution of Magpies
1 Common Magpie *Pica pica*; 2 Azure-winged Magpie *Cyanopica cyana*; 3 Green Magpie *Cissa chinensis*; 4 Yellow-billed Magpie *Pica nuttalli*; 5 Ceylon Blue Magpie *Urocissa ornata*

south to southeastern California, and as far west as western Texas and the Mississippi River in Minnesota.

The pied plumage and long wedge-shaped tail of the Common Magpie are very distinctive. The middle feathers of the tail stand out far beyond the rest, the central pair being 0.8–1.6″ (20–40mm) longer than the next pair and the rest graduated in steps of about 0.8″ (20mm). The tail feathers are black below; above both webs of the central pair and the outer webs of the rest are a brilliant bronze-green with a red-purple band near the tip. The Magpie's wings are relatively short, the first primary is scythe-shaped and very narrow, stiff and short. The scapulars, flanks and belly are white with the rump varying from white, brownish-white, brown, or sometimes nearly black. The rest of the plumage is a velvety black with the crown glossed greenish and the remainder having a blue-purple iridescence. The body is not much bigger than that of a blackbird, 10″ (25.4cm) of its 18″ (45.72cm) length is tail, although it weighs 8.5oz (240g), double a blackbird's weight. Its powerful black, slightly hooked beak, large head and strong black legs and feet make up the difference. It has strong rictal bristles, and nasal bristles completely covering the nostrils.

The Magpie's strong legs and feet, coupled with speed and agility in attack, afford it superb fighting tactics. As its enemy resists its advances it retreats with semi-hovering flight, flying backwards just out of reach, then attacks again with a quick stab of its beak and a short, sudden 'tchuk', retreating once more and returning to assail the enemy with rapid strokes. When fighting on the ground it adopts its typical excited display, which consists of flicking the wings half open and then closing them, so that they lift it forward 2ft (0.6096m) each time it hops. In one extraordinary happening at Hugglescote, Leicestershire, Magpies attacked school children, diving on them like bombers during outdoor classes. This may have been due to human predation during nesting.

Magpies are notorious thieves, taking young birds and eggs. They also eat small mammals and have been known to attack sick or injured livestock. One record tells of a group of Magpies, in hard weather, attacking the wounds of a saddle-sore donkey, eventually killing it and eating the carrion. They are one of the few birds in Britain known to attack and kill adders.

Agriculture has found Magpies beneficial because of the insects they take. They also perch on cattle and deer and rid them of ticks. Bright objects attract them. One Magpie had a liking for clothes pegs and would peck them off the line. A kindly lady placed a plastic dish of water down for it to drink but it flew away with the bright coloured dish. Tame Magpies, possibly freed by their past protectors, sometimes invade people's houses, perch on their shoulders and expect to be fed and petted. They are generally noisy and mischievous like the rest of the Crow family.

In other respects the Magpie resembles the Crow. It has the same determined strut and rather slow, direct flight, but its wing-beats are more rapid. It also hops sideways, briskly, with open wings, its long tail elevated and carried clear of the ground, often spread fanwise. It is gregarious at roosting places in woods and copses, and is known to roost in several hundreds in Europe. Like Crows, it is a frequent bather in shallow ponds and puddles and occasionally it stands with partly spread wings and ruffled plumage to catch a shower of rain. It also reacts to rain by sleeking its plumage and shaking periodically. It preens and oils its feathers after bathing, scratching its head by lifting its foot over a lowered wing in common with the Treepies and the Azure-winged Magpie, the Green Magpie and the Jays. It indulges in anting only occasionally.

On taking flight the Magpie gives a small, upward flick of the tips of its closed wings and lowers its tail slightly, but when it is alarmed it bobs rapidly, flicks its wings widely and hastily lowers its tail before rising quickly.

The noisy ceremonial assemblies of Magpies are one of the mysteries of the bird world. They occur during the first six weeks of the year and half-a-dozen to 200 birds may take part. As with the Jays, it is believed these are 'Crow marriages' which enable unpaired birds of 6–22 months to find mates, although they attract paired and unpaired birds from a wide area. Pairing is usually delayed as long as 17–22 months, but bereaved breeders may take an immature bird for a mate and even continue to nest. The 'marriage' ceremony is provoked by the forthcoming breeding season which stimulates the birds; the same assembly ground is used each year and the non-breeding birds remain for the summer. During these assemblies the

Magpies spring and dash about either in trees or on the ground with much conversational chattering; they posture and chase each other. Sometimes one bird will leave a group perched on trees and fly with slow wing-beats in a circular movement 50–100 metres, high in the air, and then alight near his fellows. During displays the Magpie's head-feathers are erected and depressed and the lifted tail is opened and closed like a fan.

Both year-old and adult Magpies defend territories in the spring. While perched on a tree Magpies fluff out the white feathers on their shoulders and sleek their black plumage. They appear to have less distinct territories than Crows; they occasionally build their nests as near as 82 metres apart. The breeders and non-breeders occupy the same territory, the non-breeders often roosting in old nests and moving further than the breeders who remain near their territory. When mobbing humans, cats, and potential predators the breeders are joined by their uncommitted relatives. Territorial holders fight among themselves and can be very aggressive, pecking at each other until one gives in. This aggression and the strain of breeding cause the greatest mortality both in adult and year-old birds in spring and early summer.

The favourite breeding territory of the Magpie is grassland with thick hedges, deciduous trees, thickets and the outskirts of woods. It is less persecuted in Europe than in Britain and can be found in open country near farms and dwellings. In Scandinavia it resides near habitations in treeless country where it nests under eaves and on telegraph poles. Where possible a breeding pair will choose a fairly tall tree so that they have a good view of their surroundings. They prefer small copses to large woods.

In courtship the pair make use of their orange nictating membrane with its blinking action, and when the female preens her partner or solicits courtship feeding she gives the begging call, a two or three syllable note: 'cheeuk-uch'. When at peace together the pair give a variety of low pitched calls. The female indicates sexual urges by giving a sudden short 'tchuk' while dipping her head, raising her tail and slightly spreading her wings. This corresponds with the clicking note made by the female *Corvus* species and some of the Jays. The pair mate once a day, early in the morning and coition takes 10–20 seconds.

Both sexes help to build the nest, the male bringing the material and the female arranging the sticks over the well of mud, and lining it with roots, grass and hair. When the female needs more material she gives a prolonged, low, husky call to attract her partner. Over the top of the nest the birds arrange an open-work dome of sticks or foliage with an entrance at the side, thus making sure the nestlings are protected from predators. Magpies often build more than one nest but they rarely use an old one. When the nest material is dropped they give a chattering call, as though in annoyance; they use this call when they are alarmed or angered, or when they are being pestered by their young for food. Alarm calls upon seeing a stranger are often 'cha-cha-cha' preceded by a longer call; when danger cannot be seen the birds give a harsh 'shrak-ak'. The female's call is similar but quieter and more high-pitched.

In April or May the female lays 5–7 eggs with an interval of 2–3 days between each egg. There is only one brood but if the clutch is lost she will lay another. The greenish-blue or greenish-grey eggs are closely mottled with brown or ash, their size (34.8 × 24.7mm) is relatively small for the Magpie's size. Incubation takes 17–18 days.

Soon after the naked nestlings hatch they clamour noisily for food, opening their beaks to reveal deep, flesh pink mouths with small white spurs at the base of the tongue and palate. They remain in the nest for 22–27 days, and when fledged still depend upon their parents for food for about 6–8 weeks, which is much longer than the Crow and accounts for their lower mortality rate. After leaving the nest they draw attention to themselves by their feeding calls. The family remain together into autumn and early winter. The young have their parents' pied plumage but the black parts are a sooty blackish-brown and the under parts buffish-white. Their wing- and tail-feathers are less glossy; the tips of the primaries are broader and their tails are shorter.

Breeding Magpies would seem to start their moult later than the other *Corvidae*, after the young cease to depend on them for food. The non-breeding, mostly one-year-olds, start to moult quite four weeks before the breeders, but their moult takes longer.

Magpies may not suffer great losses from predators; the larger birds of prey and foxes may kill some of them, but they are very agile and able to defend themselves. The Great Spotted Cuckoo,

Clamator glandarius, is a brood-parasite of Magpies in continental Europe. It is Magpie-like in some of its attitudes and lays several eggs in one Magpie's nest. The old nests of Magpies are sometimes used by Jackdaws or those agile climbers, the weasels. Where thin twigs are not available for the nest-top, Magpies build open nests which are in danger of predation by Crows and squirrels.

The Magpie's woodland origin is apparent in its method of scratching and probing the earth for grain and hidden food, its preference for feeding near thick cover, and its varied inherent calls used to contact its species in dense foliage. In common with the Crow it eats flesh and carrion, holds food with its foot, hides surplus food and tucks away bright objects. Insects are its main food and it will jump into the air to catch flying ones. It also eats cereals, nuts, fruit, peas, potatoes and berries. After feeding it hammers its bill against a trunk to clean it and strops it against perches.

The British population of Magpies is resident and sedentary, moving around only in late winter. It is very scarce in northwest

12 The Common Black-billed Magpie *Pica pica* occurs in parts of Asia, Europe, North West Africa and North America (*Eric Hosking*)

England and Scotland but common in Ireland except in the extreme west. It is also resident in Europe, but is less sedentary there than in Britain.

The nominate race *Pica pica pica* besides being found in Britain, resides in southern Scandinavia, Denmark, Poland, Germany (except the Rhineland), central and southeastern Europe, north to the Carpathians and east to Romania, Bulgaria, Thrace, Turkey, Cyprus and the Near East. In Rhineland Germany, Belgium, France, Switzerland, Italy, Sicily, Dalmatia and south to Macedonia and Greece it is replaced by *Pica pica galliae*, which usually has a dark rump. In northwestern Africa the subspecies *Pica pica mauritanica* has a black rump and a blue naked patch behind the eye. The Spanish and Portuguese subspecies *Pica pica melanotos* also has a black rump sometimes with a paler patch. *Pica pica fennorum* of Finland, northern Norway and northern Sweden across to the Baltic countries and western Russia, has a longer wing.

Pica pica bactriana, the Kashmir or white-rumped Magpie, spreads from East Russia, south through the Aral–Caspian region to Iran, south Iraq, Afghanistan and Russian Turkestan to Lake Balkhash. It is resident and subject to winter–summer altitudinal movements in the mountainous northwestern regions of West Pakistan and India. In Baluchistan its distribution coincides with juniper forests at an altitude between 6,562–8,202ft (2,000–2,500m). It also occurs in the Northwest Frontier districts between 4,921–14,763ft (1,500–4,500m). It favours such districts as Ladakh's cultivated upland valleys among the barren windswept hills and near villages with orchards and planted willow and poplar groves.

It is very similar to the Common Magpie but slightly larger with a relatively longer tail. Its first primary, which is scythe-shaped as in the Common Magpie, has a black outer and a white inner web. It has a white rump and a long, graduated black tail glossed bronze-green and purple. Its personality is corvine; it is inquisitive, wary, suspicious, and cunning if persecuted; but tame and fearless around the village houses where it is known as *Duzd* (thief).

It struts about on the ground in search of food with a rather upright carriage, tail held low, jerked and flicked at all angles in quick reaction to excitement and emotion. Sometimes it makes long hops. At sunset it makes a curious 'Ker-plonk' call as it flies with a rather laboured, flapping action, in flocks of up to thirty birds to roost communally in the trees. When alarmed it utters a loud, harsh 'kekky-kekky' call, run together in a kind of rattle, with the bill wide open, the head and neck slightly outstretched, the wings flicking and the tail jerking upwards. Another note is a more subdued, rasping 'querk'.

In the Ladakh district it breeds from March to May. Its massive domed nest is built 6.56–9.84ft (2–3m) up in willow, poplar or ilex trees near villages. The same nest is often re-used; five nests, one on top of the other making a very high pyramid have been found. In winter the nests are used as roosting shelters. For 17–18 days the female incubates the 3–7 pale blue-green eggs covered with blotches of reddish brown, (36.5 × 24.9mm), and both parents feed the young. In the first year the first primary is not scythe-shaped and the tip of the inner web is black.

Similar to the Kashmir Magpie is the Tibet or Black-rumped Magpie, *Pica pica bottanensis*. It is the largest subspecies and has the biggest bill and proportionately shortest tail, which is only just over twice the length of the wing, compared with a tail which may be up to a third longer in other races. It occurs chiefly north of the main Himalayan range in Assam, Bhutan and Sikkim, and eastwards to southeastern Tibet and northeastern China. It frequents buildings, cliffs, cultivated fields, and clumps of trees around upland villages. In addition to its harsh corvine calls it has a 'scape' or 'pench' note.

Further east from the Upper Burmese Hills through China to South Japan, North Korea, Hainan and Formosa, the Chinese Magpie, *Pica pica serica* is closely allied to the Common Magpie, but has a blue, instead of a green, gloss to its wings and rather less white plumage. It has a larger bill and much longer leg. It breeds freely in the Upper Burmese Hills in February, March and early April, and in northeast Hopeh it has been found breeding in May and June. The nest and eggs are similar to those of the Kashmir Magpie but the eggs are slightly smaller (35.5 × 24.3mm). In the Chin Hills this Magpie is host of the Koel, *Eudynamys scolopacea*, a parasitic cuckoo. It was introduced into Japan in 1598 but is restricted to northern Kyushu. It is common in the National Park and is distributed locally at Saga where it is called the 'Magpie of

Saga' and especially noted for its large nest. The Japanese have made it a National Monument.

The Black-billed Magpie, *Pica pica hudsonia*, of North America is similar but slightly larger than the European species. Its upper rump is greyish white. It has a harsh voice and gives a rapid 'yak, yak, yak'. In common with the European Magpie it has a preference for breeding in small patches of trees rather than heavy forests; it also breeds along ravines, shelter belts of trees, beside streamside shrubberies and woodland openings. The characteristic domed nest contains 6–9 greenish-grey eggs, heavily blotched brown. Incubation takes 16–18 days. The main predator of this species is the racoon.

The Black-billed Magpie's breeding range has expanded eastward and northward within the past half century. It now breeds in south and central coastal Alaska and western Canada, the interior of British Columbia and southern Yukon, south to central eastern California, Nevada, Kansas, Nebraska and South Dakota. In winter it is mainly non-migratory, but flies to the north and east, wandering to southern Mackenzie river and the Keewatin district of the Northwest Territory; also northern Saskatchewan, northern Manitoba, western Ontario and coastal British Columbia.

The other species of *Pica* in North America, the Yellow-billed Magpie, *Pica nuttalli*, is considered distinct enough to warrant specific status. This Magpie of the farming areas of California is found west of the Sierra Madre axis, chiefly in the foothills of the Sacramento and San Joaquin valleys, from Redding, Shasta County, south to Kern valley, and in the valleys among the coastal ranges from San Francisco Bay southeast to Ventura. This yellow-billed Magpie is similar but slightly smaller than the Black-billed species, and builds the same large domed nest. According to Coues (1894) *P. nuttalli* is a 'perpetuated accident' of *Pica p. hudsonia*. Yellow-billed Magpies have occurred as sports in some of the Old World races.

One of the loveliest birds in the world is the Azure-winged Blue Magpie, *Cyanopica cyana*. It has a remarkable range; the European species is confined to the west and centre of the Iberian Peninsula; a few recorded in southern France are thought to be escaped caged birds. The Asian species are found as far distant as the east temperate region of Asia and in Japan.

This beautifully coloured, graceful bird is intermediate between the two species of Magpies, *Pica* and *Urocissa*. It has the blue and white coloration with white-tipped tail feathers of the species *Urocissa* of India and the Far East; and the black bill, shaped as in *Pica*, with the same well developed nasal bristles. The Azure-winged Magpie is one third smaller than the Common Magpie, with a similar shape, but it is more delicately built, has weaker legs and silkier plumage. The bird's light grey or brownish-grey upper parts, and almost white under parts, are enhanced by its azure wings and tail and velvety black cap and face. In the Iberian race the plumage is a warmer brown above and less pale below with the tail feathers completely blue, whereas the many eastern subspecies have white-tipped tails.

In common with *Pica* it is primarily a woodland bird, favouring open woodland and country with groves of trees, thickets or orchards, riverside woods, parks and large gardens. In Europe it has a preference for warm valleys. It also resembles *Pica* in being highly sociable, gathering in small parties and flocks of around

13 Azure-winged Blue Magpie *Cyanopica cyanus* at the nest with young (*Eric Hosking*)

thirty birds. In Europe it nests in trees in small colonies about 100 yards apart with seldom more than one nest in a tree. The nest is not domed but open as in *Urocissa*. In Asia it often builds its nest in piles of driftwood lodged in riverside shrubs and bushes by the floods.

Breeding takes place in May, June or July but it has been known to occur as early as March in the Iberian Peninsula. If food is plentiful there may be two broods. The 5–9 eggs are buff, greenish or off-white with dark brown spots underlying greyish-violet markings. Like *Pica* this Magpie is host to the Great Spotted Cuckoo.

The Azure-winged Magpie has corvine feeding habits. It feeds in trees and on the ground, holds food with its foot while tearing it apart with its bill and hides surplus food. It eats insects and invertebrates, including the Colorado beetle and it also takes acorns and fruit. It has a distinctive gait which is a very quick series of hops, sometimes 'polka steps'. It has a wide variety of calls similar to *Pica*. It chatters with a loud, harsh, upwardly inflected note followed by three quick, metallic 'kyui, kyui, kyui' calls. When standing upright it gives long-drawn-out, husky, sibilant calls which appear to involve great effort. While mobbing predators it utters a long, nasal screech beginning with a guttural rattle.

The Japanese race, *Cyanopica cyana japonica* has the same variety of soft, high-pitched, whistling notes, and harsh, noisy notes as a Jay, but sweeter, such as 'gle-e-ey', 'kyiu, kyiu' or 'quey, quey'. It is not a colonial nester but nests in conifers and deciduous trees 2–6 metres high, well sheltered by thick bush, but it does have help from non-breeders during the breeding season, from May to August. It favours similar breeding habitats to its western relative. In the Kwanto plain, southeast Honshu, it is found in the City village zone and it remains in Tokyo in the summer, which is at a lower latitude than the Iberian Peninsula and very hot and sticky. Most of the Magpies migrate during the winter. This Magpie is occasionally found in central Honshu, Shikoku and northern Kyushu where it is restricted to the Saga and Fukuoko Prefectures.

The genus *Cyanopica* has eight known subspecies with *C. cyana jeholica*, of central Jehol, a questionable form probably intermediate between *C.c. stegmanni* of Manchuria and *C.c. interposita* of Shensi.

The genus *Urocissa* of India and the Far East has three species:

Urocissa ornata, *Urocissa caerulea* and *Urocissa flavirostris*.

The Ceylon Blue Magpie, *Urocissa ornata*, is somewhat intermediate between the genus *Urocissa* and *Cissa*. It resembles *Urocissa* in having a uniform colour on its head and breast, but the rich chestnut on its head, nape, upper back, breast and flight feathers differs from the colour of the other two species. In flight its long, graduated blue tail displays white tips, sub-tipped with black. Unlike the other species of *Urocissa* the Ceylon Magpie has wattled eyelids like the smaller *Cissa* species. Its eyelids are deep red, with the skin around its brown eyes somewhat paler.

The coral red bill, legs and feet of this Magpie are conspicuous as it hunts energetically for food, whether scratching around on the ground, or clinging upside down on branches to pry into crevices of bark, searching for large Sphinx moth caterpillars which infest cinchona trees. It removes the stinging hairs from one of its favourite species of caterpillars by rubbing it against mossy branches; it also favours beetles, tree frogs and fruit, giving appreciative squeaks and chirps while feeding.

This Magpie is confined to Ceylon's dense evergreen forests of the southwest Hill zone up to 6,888ft (2,100m), and the adjacent forests of the west zone down to 492ft (150m). It occasionally ventures into the local tea-gardens where it is known as the 'Kehibella'. Usually it is found in parties of six or seven in company with mixed flocks of other birds, but sometimes it lives in pairs or alone. When solitary it indulges in a quaint series of squeaks, chatterings and sucking noises with various imitations of other birds' calls. With its bill wide open it utters a jingling variety of notes such as 'chink chink' or 'cheek cheek' which are far-carrying. It has a very rasping 'crak-rakrakrak' and a plain 'whee whee' and loud, rasping notes.

From mid-January to the end of March the Ceylon Blue Magpie builds a small, crow-like nest, concealed in a low jungle tree; the nest, lined with beard lichen contains 3–5 greyish or greenish-brown eggs, profusely spotted with brown (30.5 × 22.1mm).

The range of the Yellow-billed Blue Magpie, *Urocissa flavirostris* stretches from the outer Himalayas in western Pakistan, where it takes the form of *U.f. cucullata*, to eastern Nepal where it intergrades with the Eastern Yellow-billed Blue Magpie, *U.f. flavirostris*,

14 Magpies of India
   a The Himalayan Red-billed Blue Magpie *Urocissa erythrorhyncha occipitalis*
   of the Himalayas from the Punjab to Sikkim
   b The Green Magpie *Cissa chinensis chinensis* of the Lower Himalayas, eastward to Assam, eastern Bengal and hills of Assam, Burma, northern Laos,
   Tonkin and northern Annam
   c The Ceylon Blue Magpie *Urocissa ornata* of Ceylon

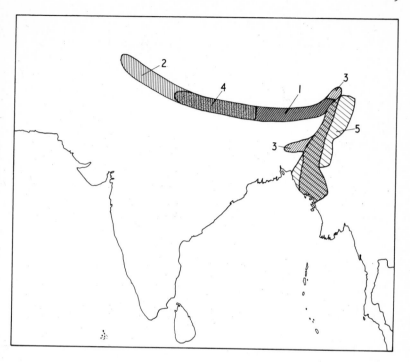

15 Distribution of Magpies
    1 Eastern Yellow-billed Blue Magpie *Urocissa f. flavirostris*
    2 Western Yellow-billed Blue Magpie *Urocissa f. cucullata*
    3 Green Magpie *Cissa chinensis*
    4 Himalayan Red-billed Blue Magpie *Urocissa erythrorhyncha occipitalis*
    5 Burmese Red-billed Blue Magpie *Urocissa erythrorhyncha magnirostris*

of the Eastern Himalayas from Sikkim as far east as northern
Yunnan. Other forms are *U.f. schaferi*, of the Chin Hills or western
Burma, and *U.f. robini* of northwestern Tonkin, North Vietnam.

The two Himalayan species are both spectacular purplish-blue
birds with black heads, necks and breasts. The eastern form has
yellow under parts and a small white patch on its nape; the western
form has whitish under parts. Both Magpies have yellow bills and
bright orange legs which are longer than those of the Ceylon Blue
Magpie. They both favour wet temperate mixed forests of pine,
oak, chestnut and rhododendron, and occupy a high zone from
1,824ft (3,300m) in summer to around 3,280ft (1,000m) in winter,
breeding mostly between 5,248–8,856ft (1,600–2,700m). In their

lower zones they associate locally with the Red-billed Blue Magpie, *Urocissa erythrorhynca*, in the west and centre of their range.

These Yellow-billed Magpies are sociable, inquisitive and rather permanently confined to their forest or wooded ravine region where they gather in noisy parties of 4–10 birds during the non-breeding season in association with the Jays and laughing thrushes. They wander slightly further afield to feed, seeking their food in low trees, and taking tree frogs, eggs and nestlings, with the parties flying from tree to tree in follow-my-leader style, their long, graduated tails fanned out displaying the black and white tips, and the streamers trailing behind. When they fly they make a few rapid wing-flaps followed by a glide, sometimes with a curious butterfly-like 'delayed action' of the wings which almost clap over the back. They venture into tea-gardens and cultivated areas and are sometimes a nuisance in hill orchards, stealing fruit, and also taking grain from harvested fields. On the ground they feed on insects and fallen fruit, hopping comically with their long tails partially cocked like a robin's.

Corvid mimicry is well advanced in these Magpies. They give loud, harsh creaking calls and sharp squealing whistles intermingled with faithful imitations of the Giant Squirrel (*Ratufa*); the Jays; the Pied Crested Cuckoo (*Clamator*); the Hawk Eagle (*Spizaetus*), and the Serpent Eagle (*Spilornis*).

Both the West and East species breed chiefly from May to June. Both sexes build the rather shallow nests, similar to but smaller than Crows', five to six metres up in such trees as a small leaf oak near the edge of a forest. Both parents incubate the 3–4 pale cream eggs, blotched bright rust, and feed the young, who upon leaving the nest are similar to their parents except that their plumage is browner and the upper parts slate brown. The white nape feathers of the parents are slightly fringed with black in the young. After a year the brown primary coverts of the immature birds are the only feature that distinguishes them from the adults.

The Formosa Blue Magpie of Taiwan Island is a separate species, *Urocissa caerula*. It lacks the wattled eyelid of the Ceylon Magpie, and has a heavier bill and longer tail than the mainland forms.

One of the most handsome members of the *Corvidae* is the Red-billed Blue Magpie, a showy, purplish-blue Magpie with a long, graduated tail, the central feathers of which are elongated into graceful undulating streamers, and a velvety black head, neck and breast. It is distinguished from the Yellow-billed races by its bright red bill and legs and its reddish-brown iris. Its range stretches from the West Himalayas around Simla and Musooree in the Punjab, east to China, and south-east to Assam, Burma and Thailand. The Magpies at the western end of the range are bluer above and whiter below than the duller northeastern species.

The Himalayan Red-billed Blue Magpie, *U.e. occipitalis*, is purplish blue with a large white patch on the nape and greyish-white under parts. Its range in the Himalayas from the Punjab to Sikkim, on hilly regions near forests, is shared with the shyer Yellow-billed species. It often visits the Himalayan hill stations to scavenge for scraps. It favours jungle and scrub in cultivated arboreal regions below 5,248ft (1,600m), and nests on wooded slopes and trees near hill cultivations.

Between April and June both sexes build the shallow open nest of twigs and tendrils and incubate 5-6 buffy stone-coloured eggs blotched brown, (33.9 × 23.9mm). Both parents also feed the nestlings. The young have a whitish crown, dark brown head and dark ashy brown upper parts, tinged bluish; under parts are whiter than in the adult. After a year their only difference is browner wings.

All the races of the Red-billed Blue Magpie have more or less similar habits to those of the Yellow-billed forms; they have the same graceful flight and glide on outstretched wings and tail; they give long hops with their tails raised clear of the ground. But they differ from the others in adopting the anting postures of the Jays, and the western form has a piercing call: 'quirer-pig-pig'.

The Burmese Red-billed Blue Magpie, *Urocissae magnirostris* of the Assam Hills to Manipur, Thailand, Cambodia, southern Vietnam, Laos and Annam, is replaced by *U.e. alticola*, in northeast Burma and northern Yunnan. These species are similar to the Himalayan population except that they have a longer bill and are darker, with more suffused deep, bright blue purple on the upper parts. They occupy the edge of the plains in tropical and subtropical dry and moist deciduous forests up to 5,248ft (1,600m) and breed in Burma from March to April.

The Chinese Red-billed Blue Magpie, *U.e. erythrorhynca* of
central and southeastern China to northern Laos and central
Annam, and *U.e. brevivexilla* of southwest Manchuria and northern
China, while having the characteristic black head, neck and breast
of the westerly forms, have a lavender patch on the nape and down
the back, with the forecrown feathers tipped with the same colour.
Their tails are azure-blue, broadly tipped with white, and except
for the long central pair, the feathers have a broad band of black
above the white tips. They have dull blue-brown wings and greyish-
white under parts, while the bluer under-tail coverts have a greyish
white band at the end tipped with black. Their breeding biology is
similar to that of the other races with the buff-coloured eggs,
heavily marked with brown being laid in May.

Another species of *Urocissa* in the Far East is *Urocissa whiteheadi*,
Whitehead's Magpie. There are two subspecies, *U.w. whiteheadi* of
Hainan Island and *U.w. xanthomelana* of Tonkin, north and central
Annam and central Laos. This Magpie has a much coarser and
more powerful bill, powerful black legs, and a shorter tail than the
other species. It is larger and has grey, black and yellow plumage
with no blue coloration. It has a black breast and paler under parts
similar to *Urocissa*, but its voice and habits resemble the Green
Magpies, *Cissa*, and it may be a link between the two genera.

The two species of *Cissa*, the Green Magpie, are to be found
in the dense mountain jungle of south-eastern Asia. The larger
species, *Cissa chinensis chinensis*, is not as large as the *Urocissa*
group, but is similar to the Himalayan Red-billed Blue Magpie in
its long tail, bright coral-red bill and legs, and the bright red ring
of wattled skin around its brownish crimson to blood-red eyes.

The Green Magpie has a striking plumage of bright lead-green
with cinnamon-red wings which have white-tipped tertials with
black subterminal bands. Its crown feathers are lengthened to form
a crest and broad black bands run through the eyes to meet at the
nape. When flying in its characteristic Treepie style it reveals its
black and white graduated tail.

The nominate species, *Cissa c. chinensis* occurs in the Lower
Himalayas, from the Jumna valley in Utta Pradesh, eastward to
Assam, Burma, and south to Tenasserim, Thailand, northern
Laos, Tonkin and northern Vietnam. The four subspecies cover

Vietnam, central Laos, the Malay States, Sumatra and northwestern Borneo.

The larger species frequents the tropical and subtropical, wet, evergreen jungles in nullahs (ravines), with dense tangles of vines; it occurs also in mixed, moist, deciduous bamboo forests at 3,936ft (1,200m), but occasionally up to 5,248ft (1,600m). It is a shy, wary bird, merging its green plumage with the foliage. Its loud, discordant, quickly repeated 'peep-peep', and raucous mewing is heard more often than the bird is seen as it flies from cover to cover. It also gives rich, melodious, squealing whistles and mimics other birds such as the Hawk Eagle.

When searching for food it is usually alone or in pairs, but sometimes it congregates in noisy parties. It associates closely with flocks of laughing thrushes, with which it shares many habits, and also with drongos; they all hunt for insects, lizards, snakes and small birds.

From April to July the birds build a well-made, massive nest with a rather shallow cup, intermixing twigs, tendrils and bamboo with moss, and making a lining of fine roots. The nest is well hidden in dense jungle, or sometimes in a clump of bamboo. The 4–6 eggs (30.2 × 22.9mm), are greyish or pale green with reddish-brown blotches, which form a cap at the broad end, similar to the Red-billed Blue Magpie's eggs; but mostly resembling those of the laughing thrushes. The young have shorter crests with paler under parts than the adults, with white tail-coverts and more pointed tail feathers.

The smaller Green Magpie, *Cissa thalassina*, is similar to the larger race but lacks the white-tipped tertials, the black bands, and has a relatively shorter tail. There are seven forms of the species, ranging from Java in the south to Hainan in the north.

The Treepies are close relations of *Pica* and have similar proportions. The genus *Dendrocitta* has the numerous stiff, somewhat short bristles which conceal the nostrils as in the Magpies; they have black bills but they are shorter, heavier and more strongly arched. Their gliding flight is similar to *Urocissa* and *Cissa*, the former being sometimes known as the Blue-pie or Occipital Blue-pie. The black legs of most of the Treepies are relatively smaller than in *Urocissa* and *Pica*. They have variegated coloured plumage

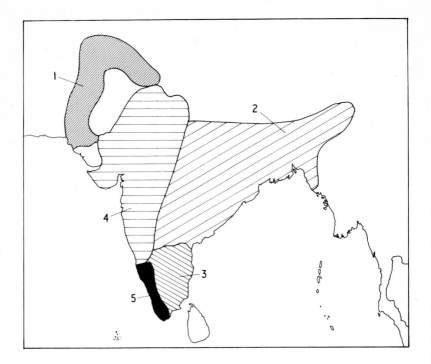

16 Distribution of Treepies
 1 Northwestern Treepie *Dendrocitta vagabunda bristoli*
 2 Northeastern Treepie *Dendrocitta vagabunda vagabunda*
 3 Southeastern Treepie *Dendrocitta vagabunda vernai*
 4 Western Treepie *Dendrocitta vagabunda pallida*
 5 Kerala Treepie *Dendrocitta vagabunda parvula*

17 Treepies (*facing page*)
 a The Indian Treepie *Dendrocitta vagabunda*, which has seven subspecies spreading from northwestern India eastwards to the Lower Himalayas, central India, eastern India, western India and southeastern India. This species is also found in western Burma and southern Burma, Thailand, Indo-China, Cambodia, Cochin-China, southern Laos and southern Annam
 b *Dendrocitta formosae*, which in India occurs as the West Himalayan Treepie and the East Himalayan Treepie, has six subspecies ranging from Burma, Tenasserim, Thailand, Andaman Islands, south China, Hainan, and the nominate race, *D.f. formosae* in Formosa
 c *Crypsirina temia* the Black Racket-tailed Treepie of southern Burma, Tenasserim, Thailand, Indo-China, Sumatra, Java and Bali
 d *Crypsirina cucullata*, the Hooded Racket-tailed Treepie of northern and central Burma
 e *Dendrocitta frontalis* the Blackbrowed Treepie of the Indian Himalayas from eastern Nepal to Assam, hills of Assam to Manipur, northern Burma and northern Tonkin
 f *Dendrocitta baileyi* the Andaman Treepie of the Andaman Islands

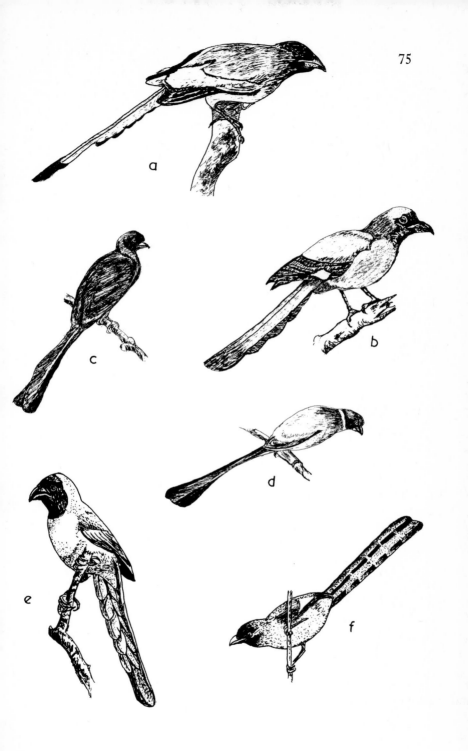

of brown, white, grey and black, and their tails are graduated, with some races having rectrices increasing in width near the tips. The other genera of Treepies, *Crypsirina* and *Temnurus* are smaller with spatulate central tail feathers and velvety frontal plumes.

Treepies are sociable and usually found in small flocks; they only descend from the trees occasionally to search for insects and small animals. Their clear, metallic voices make a pleasant sound in the jungle where they build their large, undomed nests, and lay their rather distinctive eggs, which look less corvine in most cases than the eggs of Magpies.

The range of the three genera of Treepies spreads from India to southern China, Formosa, and south to Hainan Island, Borneo, Java, Bali and Sumatra. *Dendrocitta* contains six species: *D. vagabunda* with seven subspecies; *D. occipitalis* with two subspecies, and *D. formosa* with seven subspecies. The three other species are *D. leucogastra*, *D. frontalis* and *D. baileyi*.

The nominate race of *D.v. vagabunda* is the long-tailed, bright rufous Northeastern Treepie, resident in the Lower Himalayas, central and eastern India as far south as the Godavari River. It has the characteristic dark, sooty grey head, neck and breast of the species with broad black tips to its graduated, greyish tail. When it indulges in its noisy, dipping flight, calling with a loud, harsh 'kitta kitta', it gives a few rapid wing-flaps and then sails on stiffly outspread wings and tail, revealing the greyish white and black pattern on the wings. Perched on trees with its back arched and its tail depressed and turned in under the branch, it utters a rattling 'ka-ka-ka' intermingled with a metallic, flute-like 'ko-ki-la'.

Fully endowed with corvine curiosity and cunning it will leave its wooded country and thinly spaced deciduous trees in search of winged insects, grubs, spiders and snails in the village groves, sprawling jungly gardens and compounds in urban areas, and road-side avenues; it will boldly enter verandas of bungalows to hunt wasps, geckos and small bats roosting in the roof crannies. It is highly destructive in orchards where it will substitute wild figs for the cultivated variety and also take mulberries and papaya. It enjoys foraging among cereal crops and joins barbels, green pigeons and other fruit-loving birds to feed on banyan and peepul trees.

It keeps company with racket-tailed drongos and woodpeckers

when hunting insects, and families join parties of up to twenty birds to feast on swarms of winged termites. Essentially an arboreal bird, it only hops when on the ground to pick up food with its lead-coloured bill, and keeps its long tail partly cocked. It has an uncanny sense for discovering fresh tiger kills or newly shot game animals, and travels to forest fires to feast on the fleeing refugees.

From March to July the Treepies conduct a noisy courtship and breed. The pair sit side by side, half turning and bobbing comically at each other like mechanical toys, craning their necks till the bills almost touch and uttering strange musical croaks and chuckles; they also give a long drawn out '*meeaao*'.

Both sexes build the small, crow-like nest, with a deepish cup, flimsily constructed of thin twigs and placed 19.68–26.24ft (6–8m) up in mango, salai or Acacia trees, in open deciduous forests or village groves and avenues. The 4–5 eggs are usually pale reddish-white streaked with bright rust and grey, particularly at the broad end (29 × 21.5mm). The young are brown above with buffy cream under parts. At a year old they are like the adults except for pale tips to the rectrices and browner flight feathers. The Treepie's iris is orange-brown or reddish and the dusky, lead-coloured bill gradually becomes paler towards the base.

In northern India the Northwestern Treepie, *D.v. bristoli*, is the largest of the forms. It is as richly coloured as the Northeastern race and intergrades with it in the eastern part of its range in the Lower Indus valley. It also intergrades with *D.v. pallida* in Baluchistan, the western part of its range. This Western Treepie is smaller and paler than *bristoli*, with shorter wings and tail. It is not rufous and its under parts are greyish-yellow. In the south of its range in southern Maharashtra it intergrades with *D.v. parvula*, the Kerala Treepie, which is the smallest race, and frequents the southwestern part of the Indian peninsula. It is much more richly coloured than *pallida*, with its back and scapulars dark rufous, and tawny buff under parts. Its chin, throat and breast are blackish.

D.v. parvula and *D.v. pallida* intergrade into *D.v. vernayi*, the Southeastern Treepie along its western boundary. *D.v. vernayi* is paler than *pallida* and has a sooty grey chin, throat and breast, with pale reddish-yellow under parts. It also intergrades into *vagabunda* at the Godavari delta.

There are three species of *Dendrocitta* in Burma. The Chin Hills Treepie *D.v. sclateri* is a paler version of *D.v. vagabunda*; its whole plumage appears washed out but its head has the dark grey of the species with the grey gradually merging into a pale rufous on the back.

South of the Chin Hills the Burmese Treepie, *D.v. kinneari*, ranges southward into Tenasserim and eastward into Yunnan to West Thailand. It is a much darker bird and is to be found in open country, thin forest and open parts of evergreen forest. It is tame and confiding and finds its way into civilised areas. A similar, but darker and browner bird, is the Tenasserim Treepie, *D.v. saturatior*, found in peninsula Thailand and Burma. *D.v. sakeratensis* also occurs in Thailand except in the regions inhabited by *kinneari* and *saturatior*. It is also to be found in Cambodia, Cochin-China and the south of Laos and Annam.

A less brightly plumaged Treepie without the contrasting colours of most Magpies is *Dendrocitta formosae*, a long-tailed, slender, grey, sooty-brown bird with a black forehead. In its undulating flight the white patch on its black wings and the chestnut under-tail coverts are very conspicuous. The long tail coverts are ashy grey with broad black tips and the elongated central tail feathers are blunt and rather spatulate indicating the spatulate central rectrices of the smaller *Crypsirina* group. It has a brownish-crimson iris, black bill and blackish-brown legs.

Although similar to *D. vagabunda*, *D. formosae*, the Himalayan Treepie, favours thick forest or well-wooded country in the neighbourhood of terraced civilisation up to between 600–1,500 metres, and occasionally 2,300 metres.

The West Himalayan Treepie, *D.f. occidentalis*, occurs in the Western Himalayas east to Punjab and Garhwal to Almora and West Nepal, where it intergrades with *D.f. himalayensis*, the East Himalayan Treepie, which is smaller and has a shorter tail than the western race. This Treepie frequents the Himalayas from Nepal eastwards to Assam, Burma, northwestern Yunnan, northwestern Vietnam and northern Laos. It is found in the same altitudes as the western race, in duars, duns, and the outer ranges of the mountains. In the higher elevations it favours broad-leaved forests and in duars it lives in tropical forests.

The two species have similar habits, keeping in small parties or loose flocks up to some twenty birds, often in company with laughing thrushes with which they form hunting parties. They fly around and beat their wings noisily among the foliage. They are chiefly arboreal but sometimes hop on the ground, with their tails cocked, and scratch around for insects. Besides eating the same food as *D. vagabunda* they also eat a cob-like seed covered by a hairy casing and acorns and chestnuts. When alarmed they close their wings and rocket downwards at a steep angle at each undulation. They descend on harvested, terraced fields for grain, and flutter restlessly among fruit-laden trees, flying to and fro with a noisy whirring of wings, and curiously jerky, deep, 'saw-edge' undulations, keeping up a harsh, raucous 'kokil-ko-ko-ko', and 'wokuwok awk'.

The breeding season is from April to July and during mating the Treepies give a mixture of harsh and musical songs consisting of 'tutuli-kaka', and a peculiar, rather comical, creaky, long drawn out 'plee-ee-chok'. The flimsy, corvine-type nest is built 9.8–23ft (3–7m) up a bush near scrub-covered ravines in oak forests. The 3–4 bluish or yellowish eggs are blotched dark brown (28.8 × 20.1mm). The parents make up for their conspicuous nest by being very quiet and secretive during nesting. The young resemble the adults but their darker parts are less clearly defined and the feathers have faint rufous tips. They have blackish chins and throats which merge with the dusky under parts. Their outer tail feathers are narrower than in the adult. First-year birds have duller primary coverts after the post-juvenile moult but the wings and tail of their juvenile plumage are retained.

The other six races of *Dendrocitta formosae* range from eastern India, southern Burma, Tenasserim, Thailand, Andaman Islands, South China, Formosa and Hainan.

The Burmese Hill Treepie, *D.f. assimilis*, is a darker and duller bird than the Himalayan species, with more uniform under parts, a paler brown face and throat, and a more massive bill. During summer it is confined to elevations between 1,000–6,000ft (304–1,824m), but it descends to foothills in winter.

The two races of *Dendrocitta occipitalis* of Sumatra and Borneo are thought to be the same species as *D. formosae*.

The characteristic Treepie flight of the White-bellied Treepie,

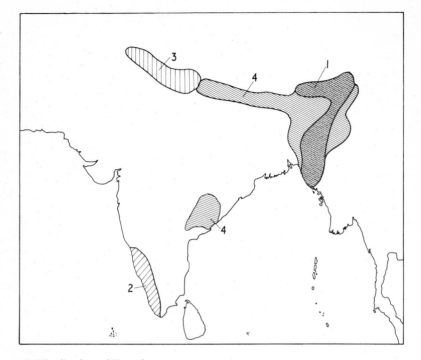

18 Distribution of Treepies
   1 Black-browed Treepie *Dendrocitta frontalis frontalis*
   2 White-bellied Treepie *Dendrocitta leucogastra*
   3 West Himalayan Treepie *Dendrocitta formosae occidentalis*
   4 East Himalayan Treepie *Dendrocitta formosae himalayensis*

Dendrocitta leucogastra reveals a large white patch on black wings
and a long, grey and black graduated tail. These features may
confuse it with *D. formosae* with whom it shares a crimson iris and
black bill. But this Treepie is the size of *D. vagabunda* and has a
black face, throat and breast, with the back of its crown and neck
pure white, together with a white belly and under parts. Its
chestnut-bay back is distinct from *D. formosae*.

   The White-bellied Treepie occupies the same southwesterly
range of the Indian peninsula as the Kerala Treepie, its range
stretching from North Kerala to Tuvandrum, and east to the
Chittoor district of Madras. Some of its habits are similar to that
species but it is shyer and shuns civilised districts, keeping to the
wet evergreen biotope, where it replaces the Kerala Treepie. The

two species generally keep apart, but occasionally both are seen in the intermediate zone where deciduous and evergreen forest types mingle. These arboreal birds keep company with parties of fly-catchers, tits, nuthatches and other insectivorous birds, particularly the racket-tailed drongos, whose calls they mimic. Their voice is similar to that of the Indian Treepie but harsher, louder and more metallic. When flitting among the trees they give a quick-repeated castanet-like rattle: 'kt-kt-kt-kt', ending in a frog-like croak. They have a humorous habit of cocking their tails with the body in a horizontal position, bobbing up and down on their perches like cheap clockwork toys.

Their courtship was thought to be observed when the mated pair were seen facing each other on the ground and croaking in a rhythmical, grotesque fashion, sounding like a heavily laden hillock cart with the wooden brakes applied and the ungreased wheels trundling up the road. They have a duck-like call during the breeding season.

Contrary to the former Treepies this species builds its crow-like nest in shrubs or saplings in heavy rain-forests away from human habitation. It has been reported to breed in February or April and again in August. The nests are built in the same place in successive years; the 3–4 eggs resemble those of *D. formosae* and are variable in colour, ranging from creamy to reddish or greenish-white streaked with brown, (28.3 × 20.5mm). The young are similar to the adults but their tail feathers are narrower and remain unmoulted in the first year; they have no rufous tips.

In eastern Nepal in the eastern Himalayas and further eastward through Assam and northern Burma to northern Vietnam, the Black-browed Treepie, *Dendrocitta frontalis*, occurs in the hills of dense, mixed evergreen forests and bamboo jungles up to 6,888ft (2,100 metres). It co-exists with the East Himalayan Treepie which it resembles except that it is less noisy and shy. It has the same reddish iris and black legs but it is slightly smaller and has a longer, entirely black tail and no black patch on its shorter wings. It has a pale grey and chestnut body plumage, a heavy bowed, black bill and similar frontal colouring to *D. frontalis*. It has a characteristic Treepie voice but a drongo-like habit of springing up in the air to catch winged termites and diving back to its perch.

Both its diet and breeding habits are similar to those of its neighbour except that it builds a more compact nest. The 3–4 bluish or yellowish eggs are more profusely marked with brown and their average size is slightly smaller (27 × 19.9mm).

A smaller Treepie, *Dendrocitta baileyi*, the Andaman Treepie, is the closest in affinity to the smaller *Crypsirina* genus. The elongated central tail feather of its long, graduated black tail increases in width towards the tip, indicative of the spatulate tail feathers of *Crypsirina*. It is a smallish, slender-bodied Treepie with the dark, bluish-ashy head, neck and mantle of *Dendrocitta*, and black feathers around the base of the bill. Its plumage would appear to merge from bluish-ashy in the upper tail-coverts and upper breast to a pale rufous on its lower back, rump and lower breast; the rest of the under parts are chestnut. The black wings have very large, white patches on the primaries and secondaries.

It occurs only on the Andaman Islands in the tall trees of dense evergreen forests. But it builds its flimsy, cup-shaped nest 15ft (4.6m) from the ground in small saplings deep in the forest. Its small, pale yellow eggs (25 × 20mm) have brown and grey blotches. Most of its habits are similar to *D. vagabunda*, but it has a sharp, metallic call sounding like a file drawn across a saw. Its iris is variable in colour, sometimes being olive-green and at others bright yellow or rich gold.

Intermediate in size between *Dendrocitta* and *Crypsirina* is the genus *Temnurus*, the Racquet-tailed Treepie, which some ornithologists have classed with *Crypsirina*; it links the two genera in the same way as *D. baileyi*. It is entirely black with the tail feathers incised along their edges, and truncated at the tips as though bites had been taken out of them. It has the coarser plumage and nasal bristles of *Dendrocitta*. Its intermediate wing length indicates the limits of *Crypsirina*. The wing length of *C. temia* is 4.65″ (118mm), that of *Temnurus* 5.39″ (137mm) and that of *D. leucogaster* 5.83″ (148mm). *Temnurus* is restricted to Vietnam and northern and central Annam and Hainan and is a little known genus.

The genus *Crypsirina* concludes the true Magpies. The two members of this genus are small with peculiarly shaped tails, the central pair of feathers being spatulate at the ends. Their black bills are short, heavy and strongly arched with the nostrils concealed by a mass of fine, velvety plumes which also surround the base of the bill. Both *Crypsirina temia* and *Crypsirina cucullata* have a blue iris. The Bronzed Treepie, *C. temia*, has a metallic, bronze-green plumage tinged with a bluish shade on the head. Its brown wings have the outer webs of the primaries greenish with the other quills glossed green. The glossy black tail has the same green tinge, its spatulate feathers measuring about 17mm across at the centre and 39mm at the tip. It has a dull black velvety forehead and face.

This Treepie frequents the open forests and bamboo-jungle of southern Burma, Tenasserim, Thailand, Sumatra, Java and Bali. It clings to outer branches of trees searching for insects and has an unusual, not unpleasant metallic call, not as harsh as other Treepies.

From April to July it breeds in a thorny bush or tree, placing its typical, though undomed, Magpie's nest 8–12ft from the ground. The 2–4 eggs are similar to other Treepies but slightly smaller (24.8 × 18.3mm).

Neither the Bronzed Treepie nor *C. cucullata*, the Hooded Racket-tailed Treepie, is gregarious, both are only seen in twos or threes. The Hooded species occurs in northern and central Burma and keeps more exclusively to the bamboo-jungle and scrub of the dry zone of Burma. It is smaller than *temia*, and is named 'Hooded' because of the ashy white ring around the neck below the black head, chin and throat. The upper plumage is vinaceous grey with the lower parts slightly more rufous. The black central tail feathers are narrower than in the Bronzed Treepie and more abruptly spatulate at the ends. The rest of the tail feathers are the same colour as the bird's back; the primaries and their wing coverts are black and the black secondaries have ashy-white edges. Breeding is confined to May. The nest and eggs (23.0 × 18.0mm) are smaller than those of the Bronzed Treepie. The young have brown heads and their central tail feathers and wings are blackish brown; their body plumage being vinaceous. The black bill of the young has an orange gape which becomes flesh-coloured in maturity. Juveniles also have pale blue eyelids with orange edges. These change to a leaden colour when the birds become adult.

Chapter 4

SOME
UNUSUAL CORVIDAE

There are three monotypic genera placed in *Corvidae* which are so distinct, and possess such unusual characteristics that it has been questioned whether they belong to the *Corvidae* family. These are the Shrike Jay, *Platylophus*, which has already been described in the chapter on Jays; the extraordinary Abyssinian Bush Crow, *Zavattariornis*, and the unique Piacpiac, *Ptilostomus afer* of Africa.

The Abyssinian Bush Crow or Pie, *Zavattariornis stresemanni*, is the most puzzling of African birds. It is about the size of a large starling and resembles this bird in its habits. Its colour and plumage pattern are also similar to some of the non-iridescent Asiatic species of *Sturnus* (Starling). It also has many characteristics of members of the *Corvidae*. It has the Magpie's bare, pigmented, bright blue area around the eye and its domed nest; the Nutcracker's long nasal bristles and long, slender bill; and its bluish-black wings and tail suggest that it derives from *Garrulus* (Jay) in a highly modified form. From anatomical studies it has been found it may be closest to the Chough with its long, slender, curved bill, long wings and moderate tail but, unlike the Chough, it has the Corvid characteristic of scuttelated tarsi. Taking all these similarities into consideration Mr C. W. Benson, an ornithologist who has studied this bird in Ethiopia, is of the opinion it belongs to the *Corvidae*, and it is so placed in Peter's Check List.

This bird, only 5½″ (14cm) long, has recently been discovered in conspicuous numbers in the park-like 'thorn acacia' country to the north and south of Yavello. It has also been noticed near Arero; in a few places between Yavello and Mega, and a few miles east of Mega. It is entirely absent to the west of Yavello where there appears to be a suitable environment and there is no obvious explanation for its remarkably restricted distribution, as there appears to be nothing unique about its chosen environment. The

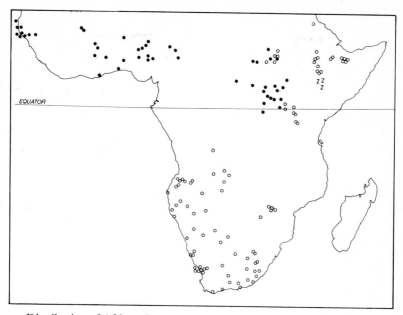

19 Distribution of African *Corvidae*
- ● Piac Piac or Black Magpie *Ptilostomus Afer*
- ○ Black Crow or Cape Crow *Corvus capensis*
- Z Abyssinian Bush Crow *Zavattariornis stresemanni*

Corvidae are poorly represented in Ethiopia; there is one Chough just holding out in the northeast; six species of *Corvus*, and the distinctive species Piacpiac.

The Bush Crow with its grey back, white under parts and black wings and tail may be seen in parties of half a dozen in acacia thorn bushes outside the breeding season from June to February, calling with a high-pitched 'chek'. March is the breeding season in Yavello district. The bird is not a colonial nester; it chooses a site at the top of a 20ft (6m) high thorn tree and builds an untidy structure of thorn twigs about 1ft (30cm) long and ⅞″ (20mm) thick lined with dried cattle-dung and dried grass. The inside of the nest is a globular cavity of about 1ft (30cm) in diameter with an entrance at the top which is protected by a vertical, tubular tunnel of about 6″ (15cm) in height, 1ft (30cm) in outside diameter, but with an internal diameter of not more than 3″ (7.5cm). This tunnel is added to the main nest just like a vertical cylinder tapering towards the top, with the entrance tunnel at the summit. Clutches of up to six eggs have been found. The cream-coloured eggs (27 × 20mm), are smooth and fairly glossy with pale blue blotches, especially at the larger end. Three birds are often seen emerging from a nest but only one female is responsible for laying the eggs.

Most of the African *Corvidae* are true Crows, but the Piacpiac, *Ptilostomus afer*, is a unique bird of uncertain affinities. It has a long, dark brown, steeply graduated tail which has earned it the common name of Magpie, but although its form is magpie-like, its black and brown plumage and its gait suggest it might be an offshoot of the *Corvus* group.

The range of the Piacpiac stretches from Senegal to Lagos, eastwards to the Sudan and southern Ethiopia, and south in East Africa to Lakes Albert and Edward. Its migratory movements are irregular.

This active, lively little bird is usually found in small parties playing follow-my-leader among the *Borassus* palms with which it is generally associated. It carries out aerial evolutions and is a gregarious, noisy bird with a deep, piping call. It also chatters and has a harsh, scolding alarm note uttered with a bob of the head. It is found in dry country with sparse trees, feeding on insects on the ground or in the palms. In the open country it is usually shy, but around camps it becomes very tame and has been said to make a good pet. It is often found in company with livestock and has the Jackdaw's habit of perching on the back of domestic animals.

In Nigeria and the Sudan it breeds from March to April. Its nest-building is crow-like; its deep cup of sticks, leaves, and palm fibre is built in the top of palm trees and the entrance is covered with thorny boughs. The 3–7 eggs (30 × 20mm) are pale blue blotched and spotted brown to pale grey.

The Piacpiac possesses the external characteristics of the *Corvidae*, such as black coloration; dense, short nasal bristles meeting above the culmen; very large black feet, and large, strongly scutellated legs; a large tenth primary, and a skull which is similar to *Corvidae* in most respects, although the vomer tends to be pointed towards the front, rather than truncate as in other *Corvidae*; and it has no lachrymals (bones near the tear-gland). It also differs from the *Corvidae* in having ten tail feathers instead of twelve. Another distinctive feature is its violet-blue or red-brown iris, and the stout, arched black bill of the adult which, in juveniles, is violet with a black tip, or in some cases pink or red with a black tip. The dark brown of the tail is also present on the wings, but their undersides are ashy. The black body has a slight gloss which is less apparent in the juveniles.

Another highly specialised genus is *Podoces*, the Ground Chough or Jay. This genus is the most un-corvine of all the *Corvidae*; it is thought to have derived from the Old World Jays, and to have evolved in response to the unusual conditions with which it had to contend in the high, barren or brushy plateaux of Central Asia. The result is a small bird of some 8″ (20cm) in length with long legs, short wings and tail, a decurved bill of long and slender proportions, sandy plumage, and the bluish-black wings and tail, and, sometimes, a dark area on the cheek, characteristic of *Garrulus*. Some of its habits are very unjaylike. It is terrestrial and fast-running, and while some species build rather jay-like nests in the bushes and lay spotted eggs, others nest in holes in the ground. This bird resembles the Nutcrackers in having a short tail and nasal bristles covering the nostrils. The two genera may have had a period of common ancestry when they diverged from Jays, but the Nutcracker, being a mountain bird, developed long wings. The long, slender, decurved bill, nasal bristles and short tail are also suggestive of the Chough.

The genus *Podoces* has four species: *P. hendersoni*, known as Henderson's Ground Jay, of Central Asia, Mongolia, northern Kansu, northern Tibet, Sinkiang, westward to Dzungaria and northwest to the region north of Lake Zaisan. *Podoces biddulphi*, Biddulph's Ground Jay, of western Sinkiang is a true *Podoces*, resembling *Podoces panderi*, Pander's Ground Jay, of the Russian Turkestan deserts. *Podoces pleskei*, Pleske's Ground Jay, is a cinnamon coloured bird the size of a hoopoe, with a black throat and tail and black and white wings; it occurs in eastern Iran and Baluchistan, entering West Pakistan near the Iran–Baluchistan boundary at Nokkundi.

Hume's Ground Chough, *Pseudopodoces humilis*, is a more distinct type which is found in southern Kansu, northwestern Szechwan, westward to Tibet, northern Sikkim and southern Sinkiang. It frequents sandy, stone-littered hillsides and hummocky country sliced by dry watercourses. It is a sprightly, active little bird, no larger than a bluebird, (*Sialia sialis*), with a greyish, sandy-brown plumage; the darker brown wings have faint terminal brown bars on the wing coverts and pale edges to the primaries. It has a whitish collar, a dark streak through the eyes and a square-ended sandy-white tail with brown centre feathers. Its black bill is long, slender and curved, its iris brown and it has brownish-black legs and claws.

This unique bird has the appearance of a rubber ball as it moves with long, bouncing hops and its body and head held erect. Every now and then it will hop onto a stone, and bob and curtsey like a chat, flicking open its wings and tail. It is to be found singly, in pairs, or in parties of six to eight, pick-axing the earth with its long bill, probing into crevices and peering under stones for food. It has a feeble flight and can fly only about 154–328ft (50–100m). It becomes tame and fearless in Tibetan villages, perching on the piled stone boundary walls and roof tops. It gives a feeble 'cheep', and a plaintive whistling 'chip' and 'cheep-cheep-cheep' quickly repeated.

In Tibet this small bird breeds at 11,000–15,000ft (3,350–4,500m) from May to July. It builds its nest in rodents' burrows or crevices in walls, making a large, untidy pad of either sheep's wool or yak hair, sometimes placed on a mossy foundation. The pad weighs as much as 700 grammes at times; it is stuffed into a widened chamber at the end of a horizontal tunnel some 75mm in diameter and 4.92–6.56ft (1.5–2m) long. The birds burrow to make this tunnel in the side of a steep earth bank or rodents' burrow. The eggs are white like those of many cavity-nesting birds; they are unmarked and have a pinkish translucence (23.6 × 16.75mm). Both sexes share in building the nest, incubating, feeding the young and cleaning the nest.

Chapter 5

NUTCRACKERS

There is a similarity between the Nutcracker *Nucifraga*, the Chough, and the Ground Jay. They all appear to be descendants of the Old World Jays, but whether they are alike enough to have a period of common ancestry is uncertain. All three genera have nasal bristles, which can be attributed to the cold climates which they inhabit. They favour high altitudes, have long bills and short tails. Here their likenesses end.

The Eurasian Nutcracker, *Nucifraga caryocatactes*, is distinguished from the other *Corvidae* in having a brown plumage handsomely bespangled with white. Its forehead and crown are dark chocolate brown and the rest of the soft, full, body plumage is paler, with pear-shaped spots at the tip of each feather. The wings are blackish-brown with the tips of the wing coverts white. Underneath the shortish, black tail, the tail coverts are white and this colour continues down the underside of the tail giving it a white border. In flight the broad wings with the fifth primary the longest, and the slightly shorter fourth and sixth primaries often equal, reveal a greater length than the straight, slightly rounded tail; and the long, black bill is seen to be about as long as the head.

There are two species of *Nucifraga*. The Eurasian *Nucifraga caryocatactes* has eight subspecies, and in North America *Nucifraga columbiana* is one distinct species.

In many respects the Nutcracker resembles the Jay. It has the same dipping flight and actions, flying freely over trees and perching on their topmost branches; but the Eurasian species is less shy and keeps less cover than the Jay. It shares that bird's preference for nuts and has its habit of burying them. Small parties of Nutcrackers often hop with Jay-like heaviness on the ground. They have similar harsh calls but these are less strident and have a greater carrying power. The harsh 'kror', and loud, rasping 'krair' are repeated fairly quickly; the birds also give a Jay-like 'skraaak' and guttural 'urkk urkk'.

Coniferous forests in mountainous regions are the Nutcracker's main habitat. It particularly favours forests of the Arolla pine. It frequents the mountains of central Europe and descends to the lowlands where there are deciduous woods and hazel coppice. It is Norway's rarest native bird and is confined to scattered localities in the southern forests. It is also a rare breeder in central and southern Sweden and south-west Finland. Its range continues across northwestern Russia, through Asia to Kamchatka, and southwards to Korea and Japan, as far inland as northern Mongolia and the Altai mountains. The more southerly Asian races extend from Russian Turkestan across the Himalayas to the mountains of northern Hopeh. There is an isolated subspecies in Formosa.

There are two distinct races of Nutcrackers in Eurasia. The Thick-billed Nutcracker, *Nucifraga c. caryocatactes*, of central and northern Europe, which is mostly limited to the mountains of Switzerland, the Carpathians and Balkans to Scandinavia, also stretching to the central Urals in Russia, the Baltic States and Poland.

The Slender-billed Nutcracker, *Nucifraga c. macrorhynchos* breeds in Siberia, northeastern Russia, the foothills of the Himalayas, south China, Korea, Japan and Formosa. Both species are rare vagrants to Britain, Denmark, Holland and Belgium and spread into southern Europe in loose flocks in winter as far south as northern Italy, Hungary, Romania and south-west Russia. The two species are indistinguishable in the field.

20 Nutcrackers (*facing page*)
    a A Nutcracker's head illustrating the throat pouch, and below the inside of the lower mandible illustrating the entrance to the pouch, which is below the tongue
    b Clark's Nutcracker *Nucifraga columbiana* of the mountains of western North America
    c The Thick-billed Nutcracker *Nucifraga caryocatactes caryocatactes* of Scandinavia, southwestern Finland, northwestern, central and eastern Russia to the central Urals, the Baltic States, Poland, southward to the mountains of western, central and southeastern Europe

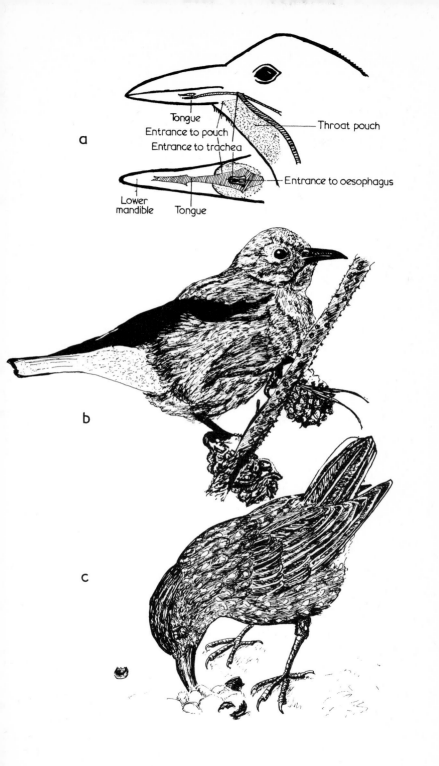

a

Tongue

Entrance to pouch

Entrance to trachea

Throat pouch

Entrance to oesophagus

Lower mandible

Tongue

b

c

The Thick-billed Nutcracker breeds in the second week in March in central Europe, and in April in the north. It heralds its mating season with a spring babbling which is not unmusical and similar to the starlings' spring song. It also gives various croaks, clicks and mewing notes and a loud: 'kerr kerr' during the breeding season. When it is actually nesting it remains silent and secretive. It is a solitary nester in conifers, close to the trunk and some 15–30ft (4.5–9m) from the ground. Its nest is built of twigs, moss and lichens solidified with earth and lined with grass and hairy lichen, *Usnea barbata*. The success of breeding depends on the hazel crop; when there is a good crop 4–5 eggs will be laid, otherwise only 3–4 bluish-green eggs with olive brown and grey marking (33.9 × 24.9mm) are laid. The survival of the young depends on the abundance of hazel nuts.

The hen incubates for eighteen days, and is fed by the male. Both parents feed the young in the nest for 3–4 weeks, bringing food in their throat pouches. During the previous autumn the adults fill their pouches with hazel-nuts and fly off to higher ground to bury them. When taking nuts from outer twigs they flutter without perching, realising the slender twigs will not take their weight. Even when the snow is 50cm deep these clever birds can detect where they have buried their supply of nuts. The Nutcracker has a projection inside the lower part of its bill which fits into the cavity in the upper part and makes the bill highly efficient for cracking nuts. After selecting a conifer cone the bird will hold it firmly in one foot and extract the seeds with its powerful bill. Besides seeds of pine, spruce, cedar and larch the Nutcracker favours the large wingless seeds of *Pinus cembra*, the Swiss stone pine, and acorns, walnuts, berries, insects and the eggs and young of small birds.

Abrasion dulls the parents' plumage as they moult from June to October. The paler brown juvenile has browner wings, less conspicuous white spots on its mantle, pale streaks across its crown, white-tipped wing coverts and usually a whitish throat. In its first year it remains less glossy than the adults and the spots wear off its greater wing coverts.

The Slender-billed Nutcracker's distinctive features are its finer, less deep, less broad bill, which is longer and more finely pointed than the former race, and its larger white tips to the tail feathers.

Although the main habitat of this Nutcracker is the Siberian conifer forests where it has a preference for the Siberian cedar, *Pinus cembra siberica*, it is not confined to conifer forests when it invades Europe in winter. It occasionally ventures as far west as the British Isles, and is often found in France, with loose flocks in the Pyrenees, northern Italy, Hungary, Romania and Russia. Its numbers vary from year to year, according to the good or bad crops of Arolla pine seeds. For centuries the Nutcracker's arrival was feared by the Ukrainians, the Poles and Germans, who looked upon the birds as omens of disaster. The Nutcrackers fly west if their pine seeds are scarce and feed on such food as hazel-nuts and insects; vagrants to Britain feed on grain and dung beetles.

Besides occurring in Siberia other races breed in Asia. *N.c. japonica* is found in the alpine zone of Japan, feeding off the pines, *Pinus punuro*; *N.c. owstoni* breeds in Formosa; *N.c. interdicta* in the mountains of northern Hopeh; *N.c. macella* in the northern Yunnan and Himalayas west to Nepal; *N.c. hemispila* in the Himalayas from western Nepal to southern Kashmir and *N.c. multipunctata* in northern Kashmir to northern Baluchistan and eastern Afghanistan.

The Siberian form builds its nest in pine forests early in the year. Its breeding habits are similar to those of the Thick-billed race. The 3–5 bluish-green eggs (33.9 × 24.9mm) are finely spotted with ash grey or pale lavender.

The Himalayas have two subspecies of Nutcracker. The Large-spotted Nutcracker, *N.c. multipunctata* of the northwestern Himalayas occupies moist and dry temperate oak and conifer forests, especially Blue pine (*Pinus excelsa*), and spruce. In Kangra, Himachal Pradesh, it intergrades with the Himalayan Nutcracker, *N.c. hemispila*. It is distinct from the latter in being a rich brown bird streaked and spotted with white, and in flight its partly spread white tail and white under-tail coverts contrast with its dark body. The Himalayan Nutcracker has smaller white spots which are lacking on the chocolate brown rump and upper-tail coverts. This form stretches from Kangra district eastwards to intergrade with *N.c. macella*, the Yunnan Nutcracker, from around east-central Nepal. It frequents the Blue pine and the spruce, prising the seeds apart from the scales before the mature cones open, or catching the seeds in its mouth from an upside down position. It also favours rhododendron, fir, and deodar forests, where they are interspersed with glades and alpine meadows. This Nutcracker collects in small flocks at a height of 6,500–7,500ft (2,000–2,300m), at the top of tall conifers; piercing the air with loud, grating calls. Sometimes it searches on the ground with long hops, breaking pine cones with vigorous hammer blows. It 'jay-flies' lazily over the tree-tops with jerky, hesitant wing-beats, but when crossing valleys it flies straight giving deliberate wing-flaps with a delayed action between each one. The tail is partially spread showing the dark central feathers in contrast with the white outer ones. These white tail feathers flash when the bird perches on a branch and flicks its tail open and shut in the shadows of the pines. The Nutcracker's rattling alarm call will assemble the flock in defence of a nesting pair.

The breeding habits of both the Himalayan forms are the same. The Large-spotted Nutcracker is thought to breed in March and the fledglings start to fly in Kashmir in late April. The more eastern races breed earlier and many young are on the wing by the end of March, when other breeding areas are still under snow. Both parents build the neat, compact nest – a platform of twigs lined with grass and pine needles – near the trunk of the tree, some 19–60ft (6–18m) up. They both incubate and feed the young. The 3–4 bluish white eggs (35 × 26mm) are speckled brown. The juvenile Large-spotted Nutcracker has pale, sandy brown upper parts, and brown wings with patchy white markings. The outer tail feathers are more pointed than in the adult and the dullish-white under parts are fringed sandy brown, except for the dull white chin, throat and under-tail coverts. Their first year plumage is less glossy and the wings browner than in adults. The fledged young follow the parents in flight uttering nasal bleats, sometimes ending in 'kraak'.

The Yunnan Nutcracker, *N.c. macella*, is resident in the eastern Himalayas from east-central Nepal southeastwards to Yunnan.

21 Distribution of Nutcrackers (facing page)
  1 European and Asian Nutcracker species *Nucifraga caryocatactes*
  2 North American or Clark's Nutcracker *Nucifraga columbiana*

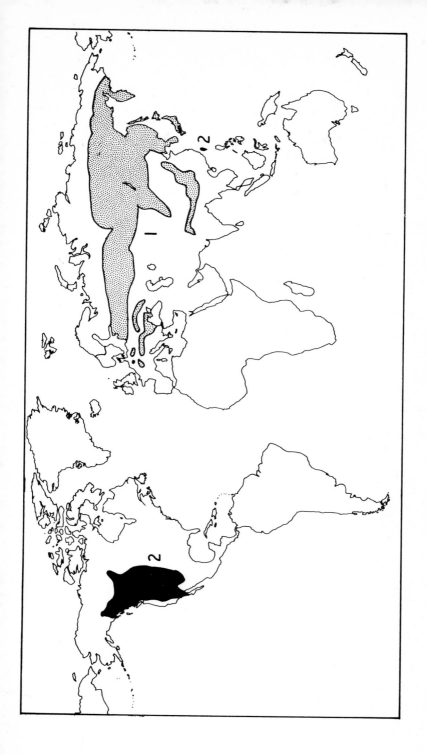

This slightly darker brown Nutcracker with an unspotted rump and less numerous white spots on the breast and back, frequents the moist-temperate and alpine conifer forests at an altitude of 6,500–11,800ft (2,000–3,600m). It is rather a weak and little known race with variable coloration of plumage.

In North America Clark's Nutcracker, *Nucifraga columbiana*, is a smaller bird than the Eurasian species. It has a more slender black bill with shorter nasal bristles. Its head, neck and body are pale, smoky grey with white about the forehead, throat and eyes. Its black wings, typical of the genus, have broad white tips to the secondaries. The central feathers of the tail are black with the rest white in tone with the under-tail coverts. It has the brown eyes and black legs of the genus. It has been compared with the Ethiopian *Zavattiornis*, and has similar eating habits to the Pinon Jay, with whom it is often in company. It is usually found in the western mountains and breeds from central interior British Columbia, east of the coast, southwards as far as eastern Wyoming and south to Baja California and western New Mexico. It is resident in Canada and wanders north to southern Yukon and eastwards to southern Manitoba. It favours open or broken coniferous woods and clearings at the higher altitudes of the western mountains. Its staple food is conifer seeds but it eats berries, seeds of lupin, larvae, butterflies, black crickets, beetles and grasshoppers. It is very fond of meat and is called the 'meat bird'; it also has a taste for suet and peanuts, and although shyer than the Jays, it will even take these foods from the hand. It is a noisy bird, its harsh, grating calls of 'charr-r-r' or 'kra-a-a' may be heard as it launches out from tree-tops, sometimes with a long swoop, opening its wings and letting itself curve up before the next drop. Its other habits are similar to those of its Eurasian cousin.

In the early spring this Nutcracker builds its nest in conifers. It makes the nest of twigs and lines it with dry grass, shredded bark, conifer needles and hair. Unlike the Eurasian Nutcrackers it does not build in any particular part of the tree. The hen lays 2–4 pale greenish eggs with pale brown or olive spots and the 16–17 days' incubation is shared by both parents. They feed their young on hulled pine seeds. The juveniles are paler and duller than the parents.

In autumn and winter this Nutcracker moves down in flocks to the lower coastal and desert regions and valleys. Irregular, spectacular eruptions of Nutcracker flocks have coincided with crops of pine seeds since 1898. In the winter they visit farmyards and camps for scraps.

Chapter 6

CHOUGHS

Choughs of the genus *Pyrrhocorax* are distinctive members of the Crow family and their relationship with other *Corvidae* is obscure. They occur in the Palaearctic region and inhabit mountainous country. There are two species: the Common Chough, *Pyrrhocorax pyrrhocorax* is a bird of seacliffs, while the Alpine Chough, *Pyrrhocorax graculus* prefers inland mountain precipices. The highest accepted altitude of any bird was made by an Alpine Chough found on Mount Everest in 1953 at 27,000ft (8,330m).

Both species of Chough are widespread over Europe and southern Asia and occupy isolated parts of northern Africa, but they are not as numerous as they were and the Common Chough in particular is facing a slow decline in the British Isles and Europe. This decline began in the British Isles during the late nineteenth century and by 1930 the Common Chough was absent from the south coast of England; it is now restricted to the southwest and west seaboard areas. It was once so common in Cornwall that it was known as the Cornish Chough but it has not bred there since 1952. In Scotland there were seventy Choughs in Argyllshire when a census was taken of their numbers in the British Isles during 1963.

In Wales the greatest number were found in Caernarvonshire (42 pairs) and the second largest in Pembrokeshire (33–36 pairs). There are about some 20 pairs in the rest of west Wales. Snowdonia can boast of Choughs among its uncommon birds. They are found in all four valleys underneath the mountains. Flocks are never seen but parties of five to six birds may be heard, if not clearly seen flying at a great height. It is thought they never nested on Snowdonia until man quarried there for they choose quarries to nest in. During the bitter winter of 1962–3 they were driven into the villages and some birds were found dead from cold. They were probably common locally rather than widespread in Britain in the past; and evidence that they did not occur in some areas is found in a monastery record at Llanelltyd, where in the seventeenth century a tame, but unknown bird like a Jackdaw but with red legs was recorded. In Penrhos Nature Reserve, Anglesey, Choughs are placed on the list of birds in danger of extinction but two pairs breed there, one pair within two miles of Penrhos. The Isle of Man has a small but stable population of 20 pairs with 6 pairs on the Calf of Man.

Ireland is the stronghold of the Chough in the British Isles. It breeds in the coastal counties from Waterford (21–26 pairs), clockwise to Antrim (31–33 pairs), except in Leitrim and Londonderry. Kerry has the largest population (132–171 pairs). There have been no Choughs on Lundy Island since 1890 attributable to Peregrine falcons forcing them too close to man.

There is no firm explanation for the Chough's decline in the British Isles and the rest of Europe. All the *Corvidae* with the exception of the Raven, have flourished in spite of man's activities. In Britain human persecution has played its part in the decline of the Chough. Large numbers of Choughs were killed by sportsmen in the late nineteenth and early twentieth centuries; and after their decline specimens and eggs were sought and Choughs were also kept in cages. The Chough is now protected under the Protection of Birds Acts, 1954–1967, and a fine of up to £25 may be imposed for wilfully disturbing a Chough while it is on the nest or has unfledged young. The Peregrine, once the chief enemy of the Chough, is now itself disappearing. Jackdaws have been held responsible for the birds' decline by occupying their territory, but Choughs prefer nesting in dark open spaces, and Jackdaws choose holes and burrows; their food preferences are also different. In some areas Choughs and Jackdaws have been found nesting together in harmony, and in a quarry the birds were reported to be nesting side by side, each using the other's site the following season. In Spain there are records of Jackdaws and Choughs competing for nesting places.

Choughs, being sedentary and also breeding in isolated communities, could be affected by in-breeding. Pseudo-tuberculosis has been suspected in some Welsh birds. Toxic chemicals must also be considered a potential threat.

Both the Common and the Alpine Choughs are black, but the Common Chough has a glossy blue-black plumage with a greenish tinge on the wings and tail, and a longer, more curved, red bill, whereas the Alpine Chough's plumage is blacker, and less shot with blue, and its bill is yellow, shorter and straighter than that of the other species. These birds are similar in proportions, colour and various habits to *Corvus*, and are in some respects intermediate between *Corvus* and *Nucifraga*. Their proportions, nasal bristles and long bills are similar to the Nutcrackers but they are very specialised with their decurved bills and broad wings with deeply slotted tips. Their red legs and feet are covered with unbroken lamina in front and behind unlike the rest of the *Corvidae*. Some of their habits are also unique.

The behaviour of the two species suggests they are closely related. They both give wing-flicking movements accompanied by a self-assertive call which is only found in the Common Magpie and the Azure-winged Magpie. Basically Choughs are gregarious; they cover the ground in flocks with long, bounding hops like Rooks; they also walk and run. Both species are normally colonial nesters but they breed solitarily where conditions are unsuitable for colonies. Non-breeders form 30% of the population in the British Isles. In 1963 700–800 breeding pairs and 400 non-breeders were found at the start of the breeding season.

In common with other *Corvidae* Choughs enjoy bathing, followed by preening, feather-oiling and scratching their heads by lifting one foot over a lowered wing. They wipe their bills with a sideways stropping action and scratch them with a foot.

The long, broad wings and broad tails of Choughs give them speed and power when flying; they have a fast and bounding flight, and their great manoeuvrability is aided by their deeply slotted wing-tips. They play in the air around cliff faces and on finding an updraught will shoot upward with head and body vertical, then flip over and plummet down with wings folded before levelling out near the ground and with a flap fly upwards again. Sometimes they turn somersaults in the air, apparently for the sheer joy of living. Their flight intention movement is a quick, upward flick of the wing-tips as they lift, and slightly open them.

Choughs are not as aggressive as other Corvids, but they defend their nest, and compete for food in flocks. The Alpine Chough being the more aggressive over food. Both species attack one another for food with head lowered and the threatened bird fluffs itself and sometimes stands upright with its bill pointing downwards, as it faces its aggressor. Later both birds may fly up and fight breast to breast. Before chasing a bird from its territory a Chough will flirt its wings and give a high-pitched 'chwee-ew'. It was the imitation of this call that gave the Chough the original name of 'Chow'. The name Chough was applied to the Jackdaw in the Middle Ages; Chaucer was probably referring to the Jackdaw when, in his 'Parliament of Fowles' he wrote of 'the thief the Chough'. The Alpine Chough gives a very different call from its relative – a high-pitched, penetrating 'chree' or 'tree', sometimes varied to 'kee', 'squee' or 'skweea'. When it is in an aggressive mood it gives a rippling 'chree-ee', accompanied by wing flirting. The alarm calls of the two species are distinctive. The Common Chough gives a harsh, scolding 'ker ker ker', while the Alpine race's call is a short, 'chrrr'. Both species have their own version of the Corvine song, consisting of low, warbling and chittering, which they give when they are peacefully feeding in flocks or pairs. They are great mimics and captive Choughs have been taught to speak.

Insects are the main food of both species. They feed mainly on the ground, probing the earth with their long bills, digging massive holes in search of insect larvae, turning over stones and pecking rapidly at swarms of ants. Alpine Choughs have been observed catching flying ants in the air. Both species hide food and cover it with stones and vegetation. They discover food discarded by picnickers in mountain passes, and the Alpine Choughs flock around ski-lifts and mountain huts where they scavenge food left by climbers. Lizards, rodents and carrion have also been recorded in the diet. Farmers have shot Choughs under the mistaken impression that they attack pregnant ewes and lambs. They are sometimes seen on sheeps' backs probing for ticks in the same way as Jackdaws. They hold their food with their feet and sometimes drink like the Jay after swallowing sticky food, but they do not dip their food in water like the true Crows.

The displays of feeding flocks of Choughs during late spring and early autumn are thought to be concerned with pair formation.

22 The Alpine Chough *Pyrrhocorax graculus* lives and breeds at very high altitudes (*Eric Hosking*)

They hop towards each other then stop abruptly with their heads lowered, their tails raised and wings flicked. Sometimes other birds attack the pair but if they are undisturbed they preen each other. On a few occasions a bird has been seen to regurgitate food which was snapped eagerly by the other bird. Choughs are not thought to breed until their second or third year, although they may pair earlier. Their aerial evolutions form a part in courtship, and preening and courtship feeding are a regular part of mating and breeding. The female adopts a begging display similar to that of the newly fledged juvenile, facing her mate and fluttering her partly opened wings; this she does when the male feeds her on the nest, but unlike other British Corvids she is often called off the nest by the male and flies half-way towards him before he feeds her; sometimes the male calls and the pair fly from the nest to join a passing flock. Solitary courtship-feeding is followed by the female squatting on the ground in the begging posture with her tail held horizontally while she quivers in anticipation of being mounted. Copulation takes from 15–30 seconds accompanied by intermittent wing-fluttering.

The Common Chough's breeding season may start as early as the end of March but in hilly country eggs are not laid until the first or second week of May. The Chough prefers inaccessible sites on ledges in dark sea caves, mine shafts and disused buildings, or in holes in crags and sea cliffs. The male builds most of the nest's outer structure out of heather, twigs and bracken, and both sexes search diligently for wool with which the female lines the nest. Although the Calf of Man has no sheep, four nests were found there lined with wool. Choughs often nest solitarily in the British Isles; nest-building takes from 2-4 weeks, then the female lays 2-6 eggs at one-day intervals or more and incubates early so that the chicks hatch out at intervals. The whitish, creamy-yellow eggs (39.4 × 27.9mm), have brownish-grey blotches and take 17-18 days to hatch. The nestlings remain in the nest for 40-45 days. The female broods them almost continuously for the first 2-3 weeks, the male feeding her and the young throughout the nesting period. Both parents feed the young before they leave the nest to wander around nearby and take their first flight.

Just after leaving the nest the young Choughs have a peculiar habit of hiding together in rocks and holes, but later they separate and hide individually. They give 'kwa-ak' calls from their various nooks to keep in contact with each other, and when they hear their parents' 'chee-ow' call they flutter from the hiding-places giving the 'kee-kee' begging call and fluttering their partly open wings. After being fed they hide until the next feeding time. They remain with their parents for 4-5 weeks, returning to the nest site to roost for several weeks; when they do leave the nest area the parents teach them to forage for food. They are like the adults only browner, with orange bills. Their legs when very young show a slight division into several scutes. For the first year they are difficult to distinguish from the adults; except for the primary coverts, secondaries and tail, which are browner than in adults; they moult their juvenile plumage from August to September. The parents moult during the breeding season, shedding their first primary in April. The moult takes 92 days.

The Common Chough has seven allied forms. A larger species, *P.p. erythrorhamphus*, is found in the Alps, Spain, Portugal, the Channel and Mediterranean Islands, except Crete. *P.p. barbarus*

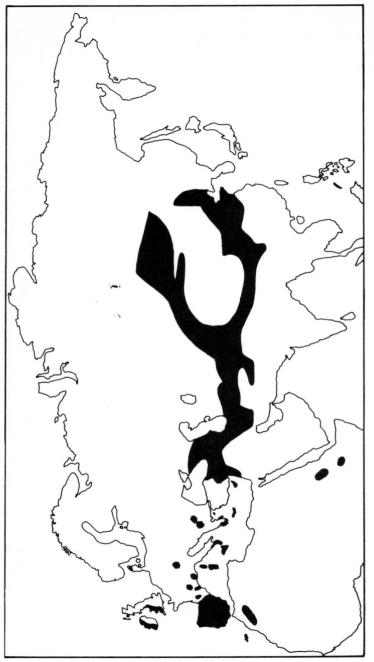

23 Distribution of Red-billed Chough *Pyrrhocorax pyrrhocorax*

occurs in the Canary Islands, (Las Palmas) and northwest Africa from Morocco to Algeria. Another larger Chough, *P.p. docilis*, with more greenish lesser and median wing-coverts, occupies Yugoslavia, Crete, Syria, eastwards to Persia and Baluchistan. In Afghanistan the population is intermediate between *P.p. docilis* and *P.p. hymalayanus*. A colony of *P.p. baileyi*, a less glossy Chough with different proportions, is found in rocky places at high altitudes on a few Ethiopian mountains on either side of the Rift, 1,500 miles (2,400km) from any other member of the species. It is thought to be a relic of the Ice Age when the whole population would have moved south.

The West Himalayan Red-billed Chough, *P.p. centralis*, coincides with the Alpine Chough except at altitudes above 9,601.2ft (3,500 metres). It is very similar to the Common Chough but larger. It occupies Russian Turkestan, its range stretching north to Baluchistan and West Pakistan, and eastwards to Kashmir and Himachel Pradesh. It frequents moist, dry temperate mountains with precipitous cliffs, steep pastures and upland cultivation, and often collects in huge flocks of several hundred during the winter to forage in fallow barley fields and yak pastures, sometimes partly under snow, in company with the Alpine Choughs, snow pigeons and sometimes Ravens. It can be destructive to ripe barley crops by thrashing out the grain from the ears of barley. It is quite tame and fearless near habitations but is not inclined to feed on the ground among houses and nomadic encampments like the Alpine Chough. Its shrill, rather musical, plaintive call of 'chiao, chiao', and high-pitched, squeaky 'khew' and korquick' can be heard when it flies at immense heights and is a mere speck in the sky.

This Chough's breeding habits are similar to those of the Common Chough. Sometimes it breeds colonially but it uses the same site each year, and from March to May are laid 3–4 white to pale salmon eggs blotched reddish or dark brown (39.2 × 27.6mm). The juvenile has a black bill with a salmon gape, tip and mouth, and black legs and feet.

East of Kashmir to Ladakh *P.p. centralis* intergrades with *P.p. himalayanus*, the East Himalayan Red-billed Chough, which in its nominal form extends from the western Himalayas, southeastern Tibet, eastwards through Sinkiang and south to the northern Yunnan to western Szechwan. This Chough has a bluish, rather

than a greenish gloss on wings and tail, which are both broader than in *centralis* with the wings more rounded. It also has longer legs. It inhabits the same climatic and geographic environment as its neighbour, living at a height of up to 19,685ft (6,000m) in mountains in summer and descending to 5,248ft (1,600m) in winter. Its habits, food and breeding are the same as *P.p. centralis*, but its eggs are somewhat larger (42.75 × 28mm).

The Alpine Chough, *Pyrrhocorax graculus graculus*, breeds at 17,000ft (5,182m), usually in colonies, in many parts of the Alps, the mountains of Spain, Morocco, Corsica, the Carpathians and the mountains of northern Iran. Its breeding habits are similar to those of the Common Chough, but the nests are built of sticks and wool in vertical cliff faces and precipices in inaccessible places. The 3–4 eggs are similar but slightly smaller than those of the Common Chough (39 × 29mm). The juvenile has a softer, browner plumage than the adult with a horny, livid bill and blackish legs and feet. It gives a curious, persistent mewing note like the chirp of a young domestic chicken.

The eastern species of the Alpine Chough, *Pyrrhocorax g. digitatus* ranges from the mountainous districts of the Lebanon to Iran, northern Baluchistan, the mountains of Russian Turkestan, eastwards to the western Altai and western Sayan mountains to the Himalayas, southern Tibet and eastern Sinkiang.

The Himalayan Yellow-billed, or Alpine Chough, is similar to its European counterpart. Its black plumage is glossed green on the wings and tail; it has short, dense nostril plumes, a slender yellow bill and bright red legs. This gregarious and sociable bird gathers in family parties and quite large flocks and occupies a higher zone than the Red-billed Chough, being more tolerant of cold. It is also tamer and more confiding and frequents upland villages, scavenging for scraps round Everest climbers' camps up to 26,247ft (8,000m) or more, and sauntering about like a starling, digging for grubs and insects.

It does not roost or nest on buildings and monasteries like the shyer Red-billed bird. It indulges in fantastic aerobatics; sometimes

24 The Red-billed Chough *Pyrrhocorax pyrrhocorax centralis* coincides with the Alpine Chough except at altitudes above 9,600ft (*Eric Hosking*)

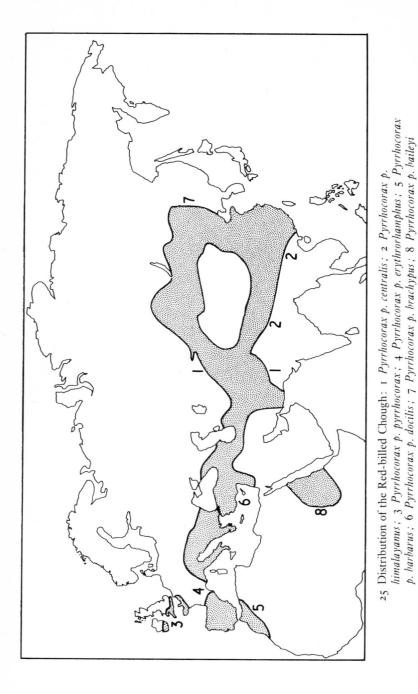

25 Distribution of the Red-billed Chough: 1 *Pyrrhocorax p. centralis*; 2 *Pyrrhocorax p. himalayanus*; 3 *Pyrrhocorax p. pyrrhocorax*; 4 *Pyrrhocorax p. erythrorhamphus*; 5 *Pyrrhocorax p. barbarus*; 6 *Pyrrhocorax p. docilis*; 7 *Pyrrhocorax p. brachypus*; 8 *Pyrrhocorax p. baileyi*

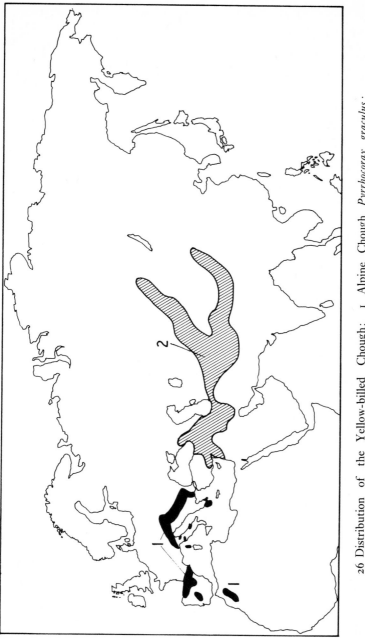

26 Distribution of the Yellow-billed Chough: 1 Alpine Chough *Pyrrhocorax graculus*; 2 Yellow-billed Chough *Pyrrhocorax graculus digitatus*

a circling flock is lifted almost vertically upward on a thermal and will nosedive with wings pulled in, several hundred metres down into valleys, ending in an effortless upward zoom, to alight gracefully on a high cliff. During its aerial frolics it gives a high-pitched, musical 'quee-ah'. It is less noisy than the Red-billed Chough but its calls are of the same type, reminiscent of the starling's cry.

This Chough's breeding habits are similar to the Red-billed's but its eggs are slightly smaller (39 × 29mm). It nests in holes in vertical cliff-faces and breeds chiefly in May and June. The juvenile has a softer, more sooty brown plumage than the adult with a horny, livid bill, olive-brown legs and feet, mottled with dark brown, and black claws. The young give the same curious, persistent mewing note as the European race.

The Australian White-winged Chough, *Corcorax melanorhampus* does not belong to the *Corvidae* family. It is a distant cousin and has the curved bill of true Choughs.

Chapter 7

CROWS

Corvus is the most advanced and successful genus of the *Corvidae*. Its large size and resilient and adaptable behaviour have enabled it to exploit many secondary habitats produced by human activities. It has spread to many areas not inhabited by other members of its group, and as a result has split into almost three times as many species as any other genus of the *Corvidae*.

The Carrion Crow, *Corvus corone corone*, is a strong and handsome bird. Upon first sight it appears to be completely black, but further inspection will reveal its glossy, greenish feathers on forehead, crown, nape and under the chin. There is a reddish-purple gloss on its mantle, back, rump, upper-tail coverts, greater, median and lesser wing coverts, and the outer webs of the secondaries. The throat, flanks, under-wing coverts, axillaries, under-tail coverts and upper parts of the tail are all glossed bluish-purple. The primaries, primary coverts and alula are black underneath and glossed purplish-green above. The under parts are black with little gloss. Throat feathers are long and lance-shaped and black forward-facing bristles shield the nostrils at the base of the thick, deep bill, which is curved at the tip and nearly as long as the head.

A Crow's flight is slow and deliberate at 20–32mph, with regular wing beats; it flies straight, hence the saying: 'As the Crow flies'. Its wings are longer than its slightly rounded tail, the fourth primary being the longest, with the second and sixth having emarginated outer webs; the sixth is slightly shorter than the third and fifth, and the first and tenth are the shortest. The secondaries are about equal in length.

Crows strut about on their large, dark feet, giving the impression they are hastening to keep an important date; occasionally they make a few hops.

Crows fostered from fledglings remain faithful to their protectors.

A boy had a Crow that slept on his bed post at night and followed
him to school. Joe, the immature Crow I fostered, was found with
tattered wings, no tail and a broken leg. He grew into a beautiful
bird and knew his name. He would eat from my hand, control my
cat and dog with his sharp beak, and spend his long summer days
'anting', collecting bright milk bottle tops and hiding them, and
bathing in his large water bowl. He would strut around carrying
brightly coloured pegs and enjoyed playing with me. I would throw
the peg and he would retrieve, or sometimes pretend to bring it to
me but then dodge out of my reach. He became so tame that I could
stroke his head and he would roll his eyes, picaninny fashion, in
pleasure.

When other Crows discovered his whereabouts they would
habitually make for his cage at 4.30am, cawing loudly. One day
they attacked him through the wire and I had to crawl into the cage
to release him, pecked and bleeding, from entangled wire. He was
in almost as bad a state as when we first found him. I bathed his
wounds and revived him with brandy. He was never quite the same
afterwards; he became nervous and only thought of escape. This

27 The Carrion Crow *Corvus corone corone* (*Eric Hosking*)

he managed twice. The first time he was attacked by roosting Crows and returned to me, the second time he achieved his aim and never returned nearer than the surrounding roof tops.

A woman reared a Crow and wanted to keep it so she clipped its wings. What had been a gay pet turned into a vicious, nervous bird, who would peck her if she came near. As we found, it is wiser and kinder to let a Crow free after necessary fostering. These birds can become a trial when domesticated. They pick up anything that interests them and have been known to fly off with five-pound notes and keys. They always approach a person from behind, and will alight on his head to inspect his ears and hair, or if a woman, her scarf, hairgrips and eventually nose and eyes.

Crows start breeding in their third year. They are fiercely territorial, and among their many communication calls is their territorial deep, hoarse 'kraah', which they proclaim from the tree-top, lowering their heads and stretching their necks to the fullest extent. This call has greater resonance than a Rook's call, with more deliberate timing and long pauses before repetition. They nest solitarily in the stout fork of a high tree, in a bush on a hillside, or a cliff ledge by the sea. Three or four pairs of Crows will nest in a favourable area, but at a good radius from each other. They will not tolerate intrusion and have been seen pecking at windows in their territory, thinking their own reflection is another Crow.

Before the trees start to bud both sexes build a large, compact nest (60cm in diameter, 10cm deep (23.6 × 3.9in)). It is a cleverly constructed bowl with a foundation of thick, dry branches broken off a tree by the male and interwoven with such strange objects as barbed wire, nails, pegs, pieces of clothing, broken china, paper, shells, sheep's bones and string, besides grass and twigs. This concoction is cemented together with mud or dung. The female lines the nest with grass, hair, fur or sheep's wool.

During courtship the male bows to the female in fairly quick succession, with his head lowered, shoulders humped, wings spread, tail fanned and moving up and down in rhythm with his head movements. Preening is an important part of their courtship. They mate once a day, early in the morning, either on the nest or in neighbouring trees, accompanied by wing-flapping by the male, and a loud, hoarse 'kaw' from the female. During coition, which takes 10–12 seconds, the female strains her tail-fan erect, clawing her perch slowly and rotating with her breast pressed downwards. All through courtship and nesting the male feeds his mate. He offers her the first nesting material as a loving hint it is time to breed, and after a while she responds and they work at building the nest in silence. At night the male leaves the female on the nest and retires to a neighbouring tree or shrub where he can keep vigil.

In April the female lays 4–5 pale bluish-green eggs speckled with greyish-brown spots (43.2 × 30.5mm). The pigment is so variable that even eggs of the same clutch may be of different shades. Incubation is by the female and takes 17–20 days. As she broods she occasionally gives a soft, crackling nest song and rarely rises except to relieve herself and stretch. Then she returns and rolls the eggs into a new position with her forehead, and separates the few feathers of her under parts so that her brood patch is in close contact with the eggs, as she lowers herself.

When the nestlings hatch they are little more than an inch of pink flesh with down on the back, short down around the inner part of the eye socket and the back of the head, a nearly naked, round, bulging belly, and eyes which are firmly closed in their gargoyle faces. Little grey legs and feet are bent helplessly under their bodies. During the first week the hen keeps them covered for the whole twenty-four hours, but later only at night. She alone feeds them at first, with regurgitated food already regurgitated and fed to her by the male. Their diet at first consists of such things as maggots and mealy worms, but later they are fed on meat of all kinds, by both parents.

The usually silent young Crows know when their devoted parents are returning to feed them and the nest becomes the scene of wide open beaks and excited chatter. The parents always approach from the same direction, alighting on the same bough and leaving the same way. Soon the parents no longer have to push the food down the nestlings' throats; the young cram their heads inside their parents' wide open beaks deep down into their crops. The fledglings remain in the nest for 30–35 days, consuming their weight in food every day and practising flight manoeuvres and wing drills before their parents, and at the same time learning the rules of the Crow community. Both young Crows and Rooks have feathered faces at this stage, but Crows are less glossed and bluer than Rook fledglings. Their mortality is greater than in Ravens, Rooks and Jackdaws; 60–70% die, although they remain near their parents and beg constantly for food. The mortality of the two-year-olds and adults reaches its peak in the early part of the breeding season. The strain of breeding and their moult weaken them. Parent Crows start to moult when they still have young in the nest. They lose four or five primaries before the secondaries and tail-feathers start to drop, and these finish about the same time as the moult of the primaries is completed. The whole moult takes 133 days. One-year-old non-breeding Crows start to moult one or two weeks earlier. In the first and second summers the young Crows have browner wings and tails than adults. Young Crows are intelligent, inquisitive and mischievous.

Crows are monogamous and stay together once they have mated. Unless they have to lay a substitute clutch they only have one brood. They can live for twelve years in captivity and probably as long in the wild.

Man is the Crow's greatest predator. Should a Crow see anyone approaching with anything resembling a gun it will sound the alarm: 'Krărr krărr krărr'. Kestrels harass Crows and some, by being persistent, manage to take possession of a nest still in use. They also use old Crows' nests.

Crows are omnivorous and have a wide variety of foods. They outnumber other *Corvidae* on farms, taking poultry, eggs and chicks and stealing grain. They kill newly-born sickly lambs, sick birds and animals, and they also rid the country of carcasses, all this being beneficial because it reduces the swarms of flies and bluebottles that multiply on sickly and dead flesh. They are enemies of Black-headed Gulls because they steal their eggs. Insects, molluscs, frogs, toads and swan mussels form part of their diet. During a hard winter they will eat potatoes and vegetables, grass, seeds and weeds but they are not often found on arable land. They pick up molluscs, crabs and nuts and drop them from a great height to break the shells. In common with Jays they bury and conceal food which they eat later.

The motorways are a source of food for Crows and Rooks. Roger Tabor, the scientist of North-East London Polytechnic, has solved

28 The Carrion Crow at nest with its young (*Eric Hosking*)

the reason why a large number of Crows and Rooks in comparison with other birds are to be found on motorway verges. These intelligent Corvines have found that the vibrations produced by large lorries, and by the out of balance loadings of empty skip trucks, bring worms to the surface and provide an ever-ready meal during normal weather conditions. They are absent during dry summers or frosty winter days, intelligently aware that worms will not surface under such conditions.

Social gatherings among Crows occur at all times and seem to be non-sexual. They come and go in pairs, chasing each other in the air and on the ground. In February parties perform sweeping, whizzing dives from great heights. During autumn and winter they gather at dusk in great numbers before retiring to a communal roost, sometimes a mile or two away. Their leader summons them with a loud, clear 'Hrrok! Hrrok!' and they manoeuvre on the ground, hopping in line down a field and turning in a methodical well-rehearsed fashion. During the day they are either solitary or in pairs, searching for food.

Fire has a fascination for them. They carry burning material to their nests and perform a characteristic anting ceremony with the nest on fire. They have a 'passive' form of anting, lying, spread-eagled, and allowing the ants to crawl over them. They are regular and inveterate bathers and play in the water by throwing in bright objects.

The British population of Carrion Crows is generally sedentary, the northern birds moving more than the southern, and the one-year-olds seeking new habitats in winter. The Dutch, French and Belgian populations are sedentary. Northern and eastern European Crows migrate up to 1,000 miles (1,600km) to their wintering grounds, but more northern populations have wintered nearer, or remained close to their breeding grounds all the year, owing to improved climatic conditions on the continent over the last few decades. As a result fewer Crows arrive on Britain's east coast in autumn.

The Carrion Crow, *Corvus corone corone*, occurs in England and Wales, southern Scotland, western Europe to western Czecho-slovakia, south to Austria, Switzerland, northern Italy and the Iberian Peninsula. Some individuals move as far south as southern Italy, Sicily and northern Morocco in winter and occasionally straggle to Madeira and the Azores.

Further east, the Eastern Carrion Crow, *C.c. orientalis*, occupies western Siberia, Hondo and southern Sakhalin, south to central Asia, Afghanistan, eastern Persia, Kashmir, Tibet and northern China. It breeds in Afghanistan and hybridises with the Hooded Crow, *Corvus corone sharpii* of Siberia, Turkestan and Persia, in a zone along the northwestern border of Afghanistan. The Eastern Crow differs from the western species in being larger, more glossy bluish-black, and having longer wings and tail, with the outer tail-feathers more graduated. It is shy and unobtrusive although it feeds in pairs wherever man has a settlement. It breeds as other Crows and nests in Chilghoza pine, chenar, willow and poplar trees.

The Hooded Crow, *Corvus corone cornix*, is the northern and eastern counterpart of the Carrion Crow and is a subspecies. Where the two interlap they interbreed and produce hybrids of inter-mediate plumage. It is understood that the two races separated

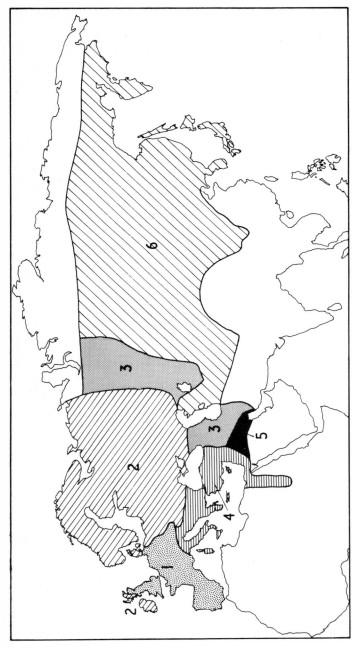

29 Distribution of the Carrion and Hooded Crows: 1 The Carrion Crow *Corvus corone corone*; 2 The Hooded Crow *Corvus corone cornix*; 3 The Eastern Hooded Crow *Corvus corone sharpii*; 4 The Mediterranean Hooded Crow *Corvus corone sardonius*; 5 The Milky-white Hooded Crow *Corvus corone capellanus*; 6 The Eastern Carrion Crow *Corvus corone orientalis*

during the Great Ice Age and upon meeting again hybridised because of their similarity. The 'Hoodie's' call and flight are identical with those of the Carrion Crow and it indulges in the same aerial acrobatics. It is more gregarious than its close relative; in the Outer Hebrides it has huge roosts of one hundred nests amongst the heather. It is resident all over Ireland and the Isle of Man. In Scotland it favours the north and northwest including the surrounding islands. Hoodies and Carrion Crows hybridise in northwest Scotland. Before 1928 the hybrid Crows tended to resemble the 'Hoodies' from a line from Kinnaird's Head southeastwards to the Isle of Man. By 1974 this line was found to have moved northwest and hybrids are found all over Caithness and half-way along the north coast of Scotland. The Carrion Crow favours lower ground than the Hoodie, and, owing to recent climatic changes in northern Scotland, its population has increased in a northwesterly direction where low ground, (under 304.8m (1,000ft)), occurs in northeast Scotland and the Moray Firth region.

I have seen large numbers of Hoodies in Denmark where they are less sedentary than the Carrion Crows. The breeding and nest-building of Hoodies are similar to the latter but the eggs are darker and some abnormally reddish eggs have been found.

The Hoodie is grey from the back of the neck to the rump and underneath from the breast to the under-tail coverts. Its black feathers have a more greenish-blue gloss. In the juveniles and immatures the backs and under parts are slightly less brownish than in the Carrion Crow. The nestlings have a little down upon hatching and the young have a greyish-blue iris which changes to dark brown.

In Finland the Hoodie is resident, but at all seasons its mortality is higher than that of the British Carrion Crow. In this land Hoodies show great initiative during winter when men fish through holes in the ice. Fishermen leave baited lines in the water to catch the fish and on their return they have found a Hoodie pulling in the line with its bill, and walking away from the hole, then putting down the line and walking back on it to stop it sliding, and pulling it again until it catches the fish on the end of the line.

Migrating Crows set their course on a fixed direction. This is illustrated by young Hooded Crows which were migrating through

30 A Hooded Crow *Corvus corone cornix* at its nest in the heather with young; it is a subspecies of the Carrion Crow and dwells in more northerly and easterly parts (*Eric Hosking*)

a bird sanctuary near Kalingrad, Russia, and were caught and taken 450 miles (750km) west to Flensburg, Germany, where they were ringed and released. Many were recaptured a long way west of their summer habitats, having continued their journey in the direction they would have followed had they been leaving the bird sanctuary after their stop-over.

An allied form of the Hooded Crow is *Corvus corone sardonius* which occurs in the Mediterranean islands, Italy, Yugoslavia, southeastern Europe, south to the Danube and the Near East, and the Nile delta as far as Aswan. It is smaller than the northern race. The subspecies *Corvus corone pallescens* of Cyprus and *C.c. minos* of Crete are paler. On Romanian rooftops and telegraph poles I found *C.c. valachus* a common sight; and it was plentiful on the shores of the Danube delta where it is renowned for its boldness in approaching human refuse and stealing the eggs of aquatic birds. The subspecies *C.c. khozaricus* of the Volga and Dona rivers,

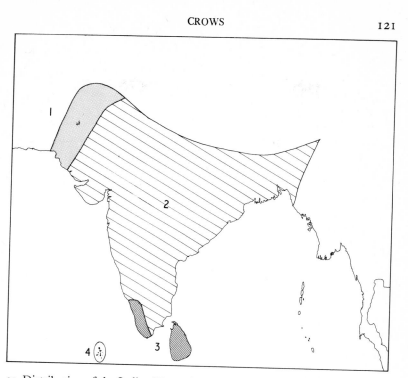

31 Distribution of the Indian House Crow
   1 *Corvus splendens zugmayeri*
   2 *Corvus splendens splendens*
   3 *Corvus splendens protegatus*
   4 *Corvus splendens maledivicus*

southern Russia, is said to have a purple rather than a green gloss
and golden-grey, rather than blue-grey plumage. From Iraq eastwards to southwestern Iran there is a much larger form, *C.c.
capellanus*, which is milky white rather than grey.

All over India except the northwest and Kerala the Indian
House Crow, *Corvus splendens splendens*, has adapted itself success-
fully to human habitation. It is a medium-sized, glossy black Crow
with a dusky grey nape, neck, upper breast and upper back; it has a
glossy black head, forehead and throat, and the wings and tail are
richly glossed green, blue and purple. It has a jackdaw-like plumage
and is more slender than *Corvus corone*. It has adapted itself completely to man's environment – probably because man over-ran its
original habitat. Before sunrise these Crows leave their roosts and
radiate in disorderly groups towards their scavenging grounds,

where, with shrill 'quak, quak', they announce their finds of kitchen scraps, garbage and carrion from gardens and stations, even hopping indoors for left-over scraps. They snap up fish stranded by receding floods and steal them from the baskets of protesting fisherwives. They raid ships as they enter harbour, and also heronries, and take eggs and sickly birds. In the semi-desert areas mice, squirrels and gerbils come under their strong beaks and they snatch at winged ants over the rooftops with ungainly aerial sallies. They are looked upon as useful scavengers and purifiers in towns and villages, and although they eat grain, fruit and groundnuts in agricultural areas this is offset by their consumption of injurious insects.

This Crow is an intelligent, inquisitive and impudently familiar bird, but wary enough to scent out danger or ill-disposed humans. It walks with a perky gait and sidelong hops, constantly flicking its wings. It has a straight, unhurried flight, and indulges in aerobatics and amusing and spectacular games. When travelling to and from its communal roost it flies very high and swoops down on its destination at a tremendous speed in a series of remarkable aerial convolutions. Besides giving the impression of reasoning, this Crow has a sense of humour, and revels in the discomfort caused by its playful tweaking at the tails of other birds, and at the ears of sleeping cows and dogs; it also pecks the toes of flying foxes as they hang sleeping in their roosts. The Crows converge in thousands at sunset to their roosts in banyan trees, mangrove jungle or coconut and forest plantations. Their clamorous cawing is mixed with the parakeets' cries and the mynas' screeching as they roost together. During the night the Crows sometimes give a strange long drawn out 'Caw' as though talking in their sleep.

The Crows pair when they are a year old and breed from October to December when they are 15–16 months, whereas the adults breed at any period from March to August. Mated Crows are very devoted; during the non-breeding season they snuggle close together on a shady branch, indulging in mutual preening and occasionally giving a contented 'kurrr'. During courtship feeding when the male regurgitates food he 'Caws' briefly and the female accepts it with a subdued 'kree-kree-kree', flirting her wings in the crouching position.

The usual untidy Corvid nest is built in the form of a branch, 3–4 metres up such trees as Mango, *Acacia arabica*, ornamental roadside trees, or on buildings. Both sexes build, framing the cup with iron wire and even gold-rimmed spectacle frames as well as sticks. The female spends hours fixing the lining of tow, coir, fibres and hair brought by the male.

The nest is 9.8–11.8in (25–30cm) wide and 2.7–3.9in (7–10cm) deep. Sometimes the Crows have three or four nests in the same tree but they are not colonial nesters. The female incubates from the first of the 4–5 bluish-green eggs, speckled brown, (37.2 × 27mm), and the male relieves her at intervals. The completely naked nestlings hatch after 16–17 days and lie with closed eyes on their delicate, almost transparent stomachs. Their flesh-coloured bodies are devoid of down until 48–72 hours after hatching; they cannot eat for the first twenty-four hours and their eyes do not open for 2–3 days. They are fed by both parents, one of which is always on guard to protect them from the hot sun and light showers. By the end of the fourth week they are fully fledged, and apart from blackish heads and throats they are deep mouse-grey. They remain with their parents for the same period as other Crows.

About 54% of nestlings survive in normal circumstances but often the nests are brood-parasitised by the Koel (*Eudynamys scolopacca*). The male cuckoo is glossy black but the female is brown with white spots above and white bars below. The feathered young Koel looks like a Crow but sometimes has a barred abdomen. This similarity is due to the Koel's art of mimicry, and the parasite does not eject the young Crows as some cuckoos do. The female Koel takes advantage of the Crows leaving the nest unguarded while they launch a furious attack on the male Koel. She lays her egg (30 × 23mm), which is greener than the Crows' but similarly speckled, and the Crows accept it. If the Crows catch the Koels in action they attack them savagely.

In thirteen days the flesh-coloured Koel hatches, blind and naked and requiring constant attention. It acquires feathers before the Crow nestlings which hatch later; and except for having its toes in two pairs, with two pointing backwards and two forward, the Koel looks like its hosts. By the time the Crow nestlings hatch the parasite is big and strong and demands most of the food so that

only one or two Crows survive. The parents continue feeding the
Koel when it leaves the nest, together with their own young.

In drier areas the forms of *C. splendens* are paler and have milky
grey necks; such as the Sind House Crow, *C. splendens zugmayeri*
of northwestern India to coastal south Iran. Those from the humid
areas of Kerala and Ceylon, *C.s. protegatus*, have very dark brown
necks. The Maldive House Crow, *C.s. maledivicus*, is similar to the
former but slightly bigger. It is abundant in the Maldive archipelago excepting Addu Atoll. Its numbers are kept down by the
Maldivian custom of shooting them on a Friday afternoon. In
common with other Indian House Crows this one is host to the
Koel parasite, but unlike them it breeds during the southwest
monsoon, between May and September.

The Burmese House Crow, *C.s. insolens*, whose range extends to
China and north of the Malay Peninsula, is similar in its audacity
and well-developed sense of self-preservation to the Indian race. Its
dark parts merge into lighter shades.

Expectation of profitable scavenging has lured the less gregarious
and sociable Jungle Crow, *Corvus macrorhynchos*, from the rural
mountainous areas of southern Asia into human habitations. It does
not have the cunning of the smaller House Crow, but it is inquisitive
and quick to learn, and is found in India and Ceylon in fairly large
gatherings where food is abundant, and becomes tame among the
tolerant people. It has widened its range by stowing away on boats
to Malaya, Africa and Arabia, where less friendly inhabitants make
it wary.

The subspecies of the Jungle Crow have differences in the bill
and body size and the feathers distinguish the different races. In
southern birds the bases of the feathers are nearly always dark; in
central India they vary from pale dirty white to dark, whilst in the
northern mountains the fully adult birds have pale and often pure
white bases to their feathers.

The Indian Jungle Crow, *Corvus macrorhynchos culminatus*, has
dusky bases to its feathers. It is the smallest subspecies and has a
relatively small but stout bill. It is found in well-wooded areas and
the outskirts of habitations all over peninsula India and Ceylon.
Besides the thievish, prankish habits of its kind, it has a peculiar
short flight with its neck outstretched, wings held vertically back to

32 The Indian House Crow *Corvus splendens splendens* has adapted itself to living alongside human habitation through most parts of India (*Eric Hosking*)

back above its body with only the tips flapping whilst it delivers high-pitched 'Caws'. Its voice is more raucous than the House Crow's but less wooden than that of the mountain Crows.

Prior to breeding the male gives gurgling chuckles and musical croaks with its head lowered, neck outstretched, bill open, throat feathers puffed out and its wing tips standing out in a point behind its rump, while it bobs its tail with each call. The female replies with short croaks, puffing her throat out in a peculiar retching motion. Because they are selective about their site, there are never more than one pair of Crows nesting in the same tree. They build their platform of sticks with a neat central cup (6–7in (15–18cm) in

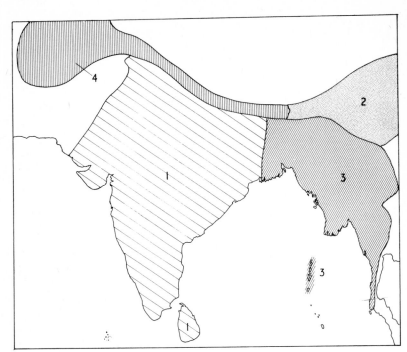

33 Distribution of the Jungle Crow
  1 Indian Jungle Crow *Corvus macrorhynchos culminatus*
  2 Tibetan Jungle Crow *Corvus macrorhynchos tibetosinensis*
  3 Eastern Jungle Crow *Corvus macrorhynchos levaillantii*
  4 Himalayan Jungle Crow *Corvus macrorhynchos intermedius*

diameter) in Mango, casuarine, *Millingtonia*, and sheesham trees. Their 3–5 eggs are similar to, but larger than those of the House Crows. The pair guard them closely and even birds of prey dare not come near them, but their aggression does not prevent the wily Koels' parasitism. When the naked nestlings open their bluish-grey eyes they also open their gape to reveal a pink tongue blotched with black. About 60% survive and are distinguishable from their parents by their shorter and browner wings and tail.

The dull black plumage of the larger Himalayan Jungle Crow, *C.m. intermedius*, has a metallic, purplish sheen, and its black bill is comparatively weak. It resembles the Carrion Crow except for its Raven-like 'Caw' and a wooden rattle-like 'kraak'. Its range stretches across the mountains from eastern Iran to Nepal in rhododendron-conifer forests, where it keeps near shepherds' encampments and hamlets. In summer it follows the sheep and goats when they are driven to upland pastures and accompanies traders' caravans to the highest Himalayan passes. It has been recorded following Everest climbers up to 21,000ft (6,400 metres). It haunts the mountain restaurants and ski-ing huts, and mountaineers can watch its spectacular aerobatics as it indulges in formation flying at a great height, each bird flying behind the other, wheeling, twisting and suddenly turning.

Except for a very large, raven-like bill, the other large Jungle Crow, *C.m. tibetosinensis* is similar to the Himalayan species in having a rather wedge-shaped tail and the same habits. Its range is from the eastern Himalayas to western China.

On the outskirts of forest villages and in towns from Bengal eastwards as far as northern Thailand to the Andaman Islands, the Eastern Jungle Crow, *C.c. levaillantii*, is distinctive with its glistening black plumage and heavy, deeply arched bill. In size it is the intermediate of the Jungle Crows and its unique call is less 'hoarse and wooden'. Its short, quickly repeated nasal 'quank quank' has been likened to Walt Disney's 'Donald Duck'. In India this race grades into *C.m. culminatus*, and in north-eastern Burma into *C.m. tibetosinensis*. In southern Burma and Thailand it grades into the nominate *C.m. macrorhynchos* of the Malay Peninsula and southern Indo-China.

In Japan the Jungle Crow, *C.m. japonensis*, is the largest Crow of that country. It is a fairly common resident around human habitations. It also breeds in the Seven Islands of Izu. Further south *C.m. connectens* and *C.m. osai* occur in central and southern Ryu Kyus, and *C.m. philippinus* in the Philippines.

In the eastern Pacific islands there are nine species of *Corvus*, of which *Corvus enca* has eight subspecies. *Corvus enca* is of smaller or medium size with a rounded wing and a long outer primary. The slightly rounded tail measures half the length of the wing. Its black plumage has a not very glossy purplish lustre; there is one small bare patch behind the eye and another at the base of the culmen. The base of the feathers are whitish on the nape, breast and abdomen.

The well-glossed *Corvus enca compilator* of Malaya, Sumatra and

34 Despite its name the Indian Jungle Crow *Corvus macrorhynchos* prefers to live on the outskirts of human habitation, in well-wooded areas (*Eric Hosking*)

Borneo is the largest form and the smallest is *Corvus enca mangoli*, a new subspecies, which occupies Sulu and the accompanying islands east of Celebes. The most glossy and purplish of these Crows is *C.e. samarensis* of Samar. *C.e. unicolor* of Banggai, and *Corvus typicus* of Celebes, are similar but the latter has white instead of black on the nape. *C.e. unicolor* is the only member of the *enca* group with white bases to the feathers; some authorities consider it to be a separate species. *C. typicus* has been reported as being unlike most Crows in its voice and habits, it is placed as a separate species from *enca*.

Although *Corvus florensis* of Flores appears to have been derived from the same stock as *enca* and *typicus*, it has a longer tail, and the basal half of the culmen and the nostrils are covered by bristly feathers; the plumage is also softer and more purplish. *Corvus kubaryi* of the Marianas, besides having the same distinctions as *C. florensis*, has feathers over the patch behind the eye.

The glossy-plumaged *Corvus validus* of the Moluccas with its steel-blue head and purplish-violet back and wings, is allied to the *enca* group, but it has a larger and longer bill and more pointed wings. The second and third primaries are longer than the first, and there is no bare patch behind the eye.

Corvus woodfordi with its powerful bill which is high at the base and curving to a sharp tip, is the only Crow to inhabit the Solomon Islands. Bristly feathers divide at the base of the culmen leaving the top of the head uncovered. This Crow and *C. woodfordi meeki* of Bougainville, Solomon Islands, have glossy, richly coloured plumage, with a greenish-bluish gloss on the head. *C.w. meeki* has a black bill covered with bristly feathers at the base.

In New Guinea and the outlying islands *Corvus tristis* stands out somewhat apart from the others because of a partial loss of head feathers and its dull plumage.

Most of the African Crows belong to the Raven group. The Black Crow, *Corvus capensis*, is the smallest of the African Corvids. It is wholly black with a long, relatively fine bill. This long bill and the fact that it has been found feeding in flocks of a thousand birds near Lake Nakuru in Kenya, have given it the name of 'Cape Rook' in East Africa and the eastern Congo, though it is more akin to the Carrion Crow. Its relationship with the latter is not regarded as close but is shown in the colour and texture of its plumage and by the fact that it breeds in single pairs and occupies a large territory. Besides the longer bill, this Crow's nasal bristles are shorter, and the breast feathers are two-forked as in the super-species, the White-necked Raven, *Corvus albicollis*, of East and South Africa. It is thought that the Black Crow may have derived from one of the larger African Ravens.

This Crow occurs in Africa from the Sudan to Somaliland, on the grasslands of Kenya at 6,000ft (1,829m), and in the near desert in southwest Africa. Its black, glossed, steely blue plumage is a common sight in East African white settlements, and although not a town bird it is found in farms and villages and replaces the Pied Crow, *Corvus albus*, in some areas in South Africa. It has a habit of perching with its head and neck feathers puffed out and giving a harsh, chuckling call. Out of the breeding season it roosts in flocks. The breeding season is irregular; in south Sudan the Crow starts to nest in December and in Ethiopia in April. In Kenya it breeds from March to July and October to December. The 2–4 salmon pink eggs have purple spots, (40 × 30mm).

The similarity between the Eurasian Carrion Crow and the American species, including the Raven, strengthens the opinion that *Corvus* reached North America from Asia. The primitive North American Crow evolved into large and small types; the large type spread into the Caribbean, differentiating into the Cuban Crow, *Corvus nasicus*, and the White-necked Crow, *Corvus leucognaphalous*, which became isolated and developed into forest types. From the same ancestral race smaller Crows evolved and spread along the coast to Mexico and the Caribbean Islands. The Fish Crow, *Corvus ossifragus*, has remained near the coast but the Mexican Crow, *Corvus imparatus*, has deviated and is the nearest Crow to the barriers of the tropical and subtropical forests of Central and South America. This Crow does not venture into the forests but occupies the semi-arid uplands.

The two Crows, the Palm Crow, *Corvus palmarum* of Cuba and the Dominican Republic, and the Jamaican Crow, *Corvus jamaicensis*, are the forest types of today. The most striking Jamaican and Mexican Crows are furthest from their North American ancestor and they have differentiated more than the other Crows. There are large and small species of Crows on all the Caribbean islands where the species occurs, except on Jamaica. This is also the case with the larger Common Crow, *Corvus brachyrhynchos*, and the smaller Fish Crow of North America, which gives rise to the suggestion that there was more than one ancestor.

There are seven species of American Crows. The large ones are the Common Crow, the White-necked Crow, and the Cuban Crow. Smaller in size are the Fish Crow, the Palm Crow, the Mexican Crow and the Jamaican Crow.

The Carrion Crow and the American Common Crow are very

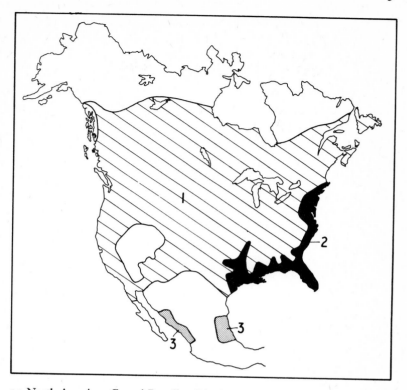

35 North American Crows' Breeding Distribution
  1 The Common Crow *Corvus brachyrhynchos*
  2 The Fish Crow *Corvus ossifragus*
  3 The Mexican Crow *Corvus imparatus*

closely related. The American Crow's length varies from 17–21″ (43–53cm), but the Carrion Crow's length is more constant, 18.5–19″ (47–48cm). Both Crows have all black plumage, the American Crow's being a metallic violet. The head of the latter is flecked with violet-blue and it has a metallic violet plumage from the upper back to the lower, with a scaly appearance caused by barbules failing to interlock. Towards the distal end the feathers grow shorter and the barbicels are either less or absent, thus producing a frayed appearance at the end of the feather.

In common with its Eurasian relative the American Crow has many calls, but its loud 'Caah', descending in pitch and varying in speed is distinctive.

The Common Crow differentiates into various species from British Columbia across to central Quebec and southern Newfoundland, south to northern Baja California and across to the Gulf of Mexico and southern Florida, generally avoiding the Great Basin. Unlike most birds the Common Crow flourishes in agricultural and coastal areas. It is a highly adaptable and resourceful bird and capable of sustaining itself in a wide variety of habitats, finding space for foraging in the broad cultivated fields, with roosting and nesting quarters in the wooded areas. Along the coast it avoids heavy forests but frequents the rivers, lakes, marshes and meadows. During the winters of deep snow it gathers round refuse dumps, searches cornfields, feeding on stubble and insects and hunts the shores for carrion and molluscs; but although it is omnivorous, its intake over the whole year is almost three-quarters vegetable.

This Crow has a stratagem for obtaining its favourite foods. It has been known to ride on a pig while it is foraging for mice. On finding a mouse the Crow gobbles it up and flies off with a joyful chuckle. It will disturb a fox slinking off with its prey by giving a great and noisy call to summon its brethren to pester the fox until he drops his prize. A pet Crow of a Long Island naturalist consumed a pint of house paint and survived! In Stewart, Ohio, a lady owned a Crow that was so dainty he insisted upon washing worms before eating them. It is a typical corvine habit to wash food.

In the latitude of the Great Lakes the spring migration of Crows is one of the most stirring events of the year. Long, loose skeins of these birds darken the sky over the lake shores and barren fields, fighting against the wind with their great wings. On arriving at their breeding areas they stay together to feed, but although they are normally gregarious they break up into pairs and remain solitary for the nesting period even if 50–60 nests occupy a small area. They sometimes become involved in the eternal triangle, and a trio may settle down peacefully, two males feeding the same brood of nestlings or two females alternately incubating one clutch of eggs and accepting food from the same male. Common Crows also steal each others' nest material.

The well-made nest is of a coarse structure hidden in a tree from 10–75ft (3–23m) from the ground; the warm yellow lining is made from the soft inner bark of chestnut, basswood or such. Both parents incubate the 4–6 greenish eggs, spotted brown, for 17–20 days. The young are fed on insects and invertebrates; when they are fledged they are dull greyish-black with only a slight violet gloss on the wings and tail; the one-year-olds are similar to the adults but have duller plumage.

The young Crows enjoy play. One will hide in a hollow tree and give loud distress 'Caws'. The flock then rush to the spot but not finding him, fly away. The Crow will repeat this many times before he pops out of hiding and 'Caws' merrily, with the indulgent flock joining in. A young Crow will collect a white pebble or shell and hold it in his beak, flying from tree to tree with the others buffeting him to make him drop it and when another snatches it the buffeting is turned on him. The adults also spend much time in practical jokes, dropping down on sleeping rabbits and rapping them over the skull or settling on drowsy cattle and startling them.

Common Crows are given to territorial mobbing of owls, hawks, racoons and foxes. Also cats – when I was on an island in the Georgian Bay I was awakened by a loud 'Caa-a-a-ing' and went to investigate. I found the island's pair of Crows mobbing the visiting cat we had brought with us to the island. The frightened creature crept under the bungalow, and the Crows, seeing me, left with angry 'Caas'.

After the breeding period Crows do not take exception to hawks and owls taking over their nests.

Summer parties of Crows merge into large flocks during the autumn and they fly to their winter roosts, sometimes 1,500 miles (2,400km) from their breeding area. As many as 230,000 Crows have been seen at a roost near Baltimore, and 100,000 at a roost near Peru, Indiana. Some roosts in these regions have been occupied since the advent of man. Owls sometimes raid Crows' roosts at night and kill sleeping birds, and during the daylight Crows will mob owls if they can locate them. They take up temporary residence near the breeding colonies of herons and gulls; but only if humans frighten the incubating birds from their nests during brooding will the Crows quietly pounce and steal the eggs.

An eminent American preacher and lecturer, Henry Ward Beecher, (1813–87), said that if human beings wore wings and feathers, very few would be clever enough to be Crows. Although bird watchers believe they know the Crow, they frequently have cause to be astonished by this feathered genius. Although the *Corvidae* have been protected during the breeding season under the Migration Bird Treaty, American local governments have offered bounties on the Crow, but in spite of the best organised and largest scale onslaughts upon him, this wily brigand continues to thrive. Few native animals in North America have been so fiercely persecuted as the Common Crow; killers bombed their roosts with dynamite in the winter and used guns and poisoned baits earlier in the year. Some 'sportsmen' have achieved a certain notoriety by inventing sophisticated methods of attracting Crows to their guns, even resorting to tape-recorded Crow assembly calls. These senseless, offensive measures against the Crow are taken in the name of 'control'. Undeserved allegations are made against the Crow by duck hunters, who claim that it has a bad effect on ducks. Farmers, who used to hate Crows, now seem to be realising flocks of Crows can be beneficial in eradicating crop-destroying insects.

Ornithologists believe that there are more Crows in the United States now than when the Pilgrim Fathers landed. The birds are almost contemptuous of human beings. Crow sentinels are posted in high trees to keep guard while the flock feeds. They can spot a gun barrel half a mile away and upon hearing the sentinel's danger signal the flock will depart silently at 45mph.

The instinctive cunning and prankish glee of a pet Crow can be a source of great amusement. William Crowder, a Mississippi naturalist, had a Crow that played with live matches; he substituted a box of safety matches. On his return home one day he found the Crow hurling matches out of the window and crying 'Ha! ha!' Another Crow, kept by a Georgian spirit-smuggler as a look-out against Revenue Officers, would mutter 'Oh boy – oh boy!' as he ogled the attractive ladies passing by. Pick-pocketing was the favourite pastime of a Staten Island, New York, Crow; when he found an empty pocket he would cry, 'Go to hell!' and fly away in a huff.

As has been indicated, mimicry is an art of the Crow. It can imitate the squawk of a hen, the crow of a rooster, and the whine of a dog. It will use this to advantage as has been found when a Crow imitated a hen to lure her from her chicks; upon failing, he walked

36 There are four races of the American Common Crow. This *Corvus brachyr-hynchos brachyrhynchos* belongs to the eastern population, extending from Newfoundland south to the Ohio River (*H. H. T. Jackson, Bureau of Sport Fisheries & Wildlife*)

maddeningly up and down in front of her until she rushed at him. While dodging her, two of his cronies swooped down and stole a chick apiece. There are records of Crows picking up farmer's expressions such as 'Giddy up!' 'Whoa!' and 'Hey!' and young Crows in captivity have been known to learn 100 words and 50 complete sentences.

There are four races of the Common Crow in America. *Corvus brachyrhynchos brachyrhynchos* is the eastern population extending from Newfoundland westward to northeastern Alberta to about 100° west longitude in the United States and southward to approximately the Ohio River where it intergrades with *C.b. paulus*, which only differs from the nominate species in the adult male's wings being shorter. *C.b. paulus* occupies the region from the Ohio River southward to the Gulf of Mexico and the northern border of Florida. Westward it ranges to eastern Texas and Arkansas.

In the Florida peninsula, the Florida Crow, *C.b. pascuus* is smaller than the nominate species, except for its larger bill and feet which are significantly larger than those of any other Crow.

The western population of *C.b. hesperis* occupies southwestern Canada and the western United States, southward to approximately the Mexican border; it is a decidedly smaller bird with a relatively smaller and more slender bill.

The Northwestern Crow, *Corvus caurinus*, was described as a distinct species by Baird in 1958, but upon careful study of its voice, habitat and measurements it would appear to be a subspecies of the Common Crow. It occupies Alaska and British Columbia and its range extends as far south as Puget Sound, Washington, where it is difficult to define the range of *C. caurinus* ends and that of *C.b. hesperus* takes over. This northern Crow is smaller than the Common Crow, with smaller legs and shorter wings which beat faster than its neighbour's. Significant differences may be found in the wings and tails of male *caurinus* between coastal British Columbia and northwestern Washington, those of the latter being larger, while the legs of the northwestern Washington Crow are shorter than those of the southwestern Washington species. Except for the legs, most measurements are similar in both sexes in south and northwestern Washington and California and Oregon, although specimens from the interior of Washington, British Columbia and Idaho are closer to the standard measurement of the Crow of California and Oregon. It is therefore difficult to make a criterion.

In Alaska and British Columbia the Northwestern Crow is essentially a coastal type because the mountain ranges and dense forests along this coast are a formidable barrier against flight inland. Crows are to be found around much of the Puget Sound, Washington, and southward to the Columbia river, where wide belts of land between the coast and the mountains have been used for agricultural farming. The harsh, low-pitched 'grar-r' of the Crows of these areas are harsher than that of the Common Crow; the birds also have a higher-pitched, more rapid call, similar to the Common Crow, but they are not considered to be pure stock. Further south the Crows' voices are of an intermediate type, *caurinus* of the northern coast intergrading with *hesperis*.

Colonial nesting is a particular feature of *C. caurinus*. The nest and eggs are similar to those of the Common Crow. From Alaska south to the Olympic Peninsula, the Northwestern Crow may be found walking around Indian villages or on the lawns of urban areas, showing little fear of the inhabitants. The reason it is tamer than its cousin is because it is not persecuted so greatly. The Indians of the Pacific northwest regard Crows as sacred birds and rarely molest them.

The Common Crow and the Fish Crow occupy the same geographical and ecological area on the Atlantic and Gulf coasts of the United States from Rhode Island southward to Florida; along the coast of Louisiana to southeastern Texas and northward following the Mississippi river basin to southwestern Tennessee. The Fish Crow has been described as an inhabitant of ocean beaches, river valleys and lake shores, but recently it has extended its range not only northward but to drier regions. It has been found in pine forests, pecan (hickory) orchards and in dry, old fields and abandoned farmlands where natural grasses and pines have taken the place of cultivated land. It depends less on agriculture for food than its neighbour and can fare adequately on natural vegetable and animal material such as berries, fruits, insects and carrion.

The Fish Crow is smaller than the Common Crow, has a more glossy violet plumage and very little scalation on its back. Its under parts are more bluish-violet than black, and its tail is much shorter than its long, pointed wings; it has stout feet and a compressed bill which is more high than broad. Its call is a short, nasal 'Car', sometimes 'cā-hā' or 'cārk', and it sounds as though it is very hoarse. The only time it is indistinguishable from the Common Crow is in the summer when the young Common Crows have somewhat similar calls.

The breeding season is from April to May, according to location. Pines are the favourite nesting sites, and this Crow builds its nest up to 90ft (27m) in the northeastern states and from 20–150ft (6–45m) in the south, a good deal higher than the Common Crow. The nests, which are in small colonies of two or three pairs fairly close together, are similar to those of the Common Crow, as are their smaller eggs.

Along the coast the Fish Crow feeds on dead animal life that floats ashore. It treads for clams as men do, dislodging them with its beak, breaking the shell and tearing out the clam with its claws. It steals eggs from heronries, and also from Ibis in the south; and in North Carolina a colony of twenty pairs of Little Blue Herons were disturbed from their nests on the Big Lake by photographers, and Fish Crows plundered every nest. They dive upon schools of fish, swallowing small fish on the spot, but bearing large ones away to tear them to pieces.

The distribution of the Mexican Crow has been the subject of speculation. On the eastern seaboard it ranges from Nuevo Leon and 60 miles (96.5km) south to Vera Cruz, and inland to San Luis Potosi. Its western range is from the Yaqin river southwards to Nayant. It has been recorded in Maria Madre Island and Colima. Although their eastern and western habitats are isolated, Mexican Crows are so similar in character that they are represented as subspecific populations. The eastern Crow is known as *Corvus imparatus imparatus*, and the western race as *Corvus imparatus sinaloae*.

The small Mexican Crow is closest in size to the Palm Crow and it is the most highly coloured of all the North American Crows; its glossy black plumage has a strong purplish sheen, with no scalation effect. It has a seasonal, gradual colour change; birds of both sexes are found to be distinctly purplish between November and early March, but greenish from April to June. This change in colour is related to chemical and/or physical change, and is found to a lesser degree in the Palm Crow, whose strikingly iridescent plumage of the autumn fades by April.

Concentrations of western Mexican Crows gather along the river estuaries and on the shore when the tide is out; they also favour coastal towns and villages and the semi-desert, deciduous woods further inland. The eastern Crows occupy semi-desert brushland, including inhabited areas and farms, but they avoid tall forests, deserts and the sea beaches. Most of these Crows are found at an elevation of between 100–1,000ft (30–300m).

Essentially an upland species, the Palm Crow of the western half of Cuba and the interior of Hispaniola, is darker and duller than the Mexican Crow. Its numbers have decreased in Cuba since the turn of the century and only a few small bands are left in arid hills scattered with pines, or around farms. It is mostly to be found in the

37 The Fish Crow *Corvus ossifragus* occupies much of the same geographical location as the Common Crow but is smaller with a more glossy plumage (*Luther C. Goldman, Bureau of Sport Fisheries & Wildlife*)

pine forests of Hispaniola; the hilly, sparsely vegetated country around La Mata; the swamps along the Samana Bay, and also bushy and arid regions, river sloughs, and dry, hot forests. It co-exists with the White-necked Crow in Hispaniola, except in the pine forests of La Salle, where the latter is absent. The Cuban Palm Crow is found with the Cuban Crow.

The Palm Crow's harsh, nasal 'Caw' has been likened to the voices of the North American Crows and the Carrion Crow. Its

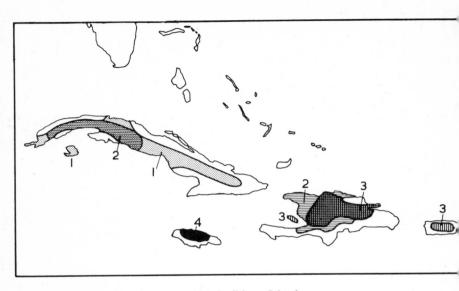

38 Distribution of Crows on the Caribbean Islands
  1 Cuban Crow *Corvus nasicus*
  2 Palm Crow *Corvus palmarum*
  3 White-necked Crow *Corvus Leucognaphalus*
  4 Jamaican Crow *Corvus jamaicensis*

raucous 'Craa-craa' resembles the Fish Crow's voice. The bird
itself is similar to this Crow from the rear view, although the Palm
Crow is duller with a suggestion of scalation; it is duller on the
under parts as well as the back. It also differs from the Fish Crow in
having longer legs, a bill with a heavier base and shorter wings and
tail.

The Cuban Crow, in common with the Palm Crow, is a disappearing bird. It has been persecuted by man and much of its
natural habitat of heavy forest has been cut down, leaving the
species abundant only in favourable areas. It also occurs in the Isle
of Pines, Grand Caicos and the Bahamas.

It is similar to the Common Crow with less scalation on the back;
the nape and back of the head are a glossy shade of deep violet. It
has glossier under parts, bare nostrils, a longer bill and shorter tail,
wings and legs than the Common Crow. It babbles and chatters and
has a variety of guttural raven-like sounds.

The Jamaican Crow has the same babbling and raven-like

croaking; it is known as the Jabbering or Chattering Crow locally and is very garrulous with its harsh 'craa-craa'. Some of its notes resemble the Palm Crow's and others are not unlike those of the European Rook. Many specimens share a resemblance with the Rook in having a bare area underneath and behind the eye and at the angle of the bill, although others are feathered.

This unique Crow with an uncrowlike appearance favours the wild wooded hills and fairly open country of north and central Jamaica. It is the smallest of the Caribbean Crows with the exception of the Mexican Crow; it is also the dullest, being sooty black with little gloss except a slight violet tinge on the wings, tail and neck. It has no scalation.

In Hispaniola the White-necked Crow shares its mountainous forested terrain with the Palm Crow. It shows a preference for inaccessible forests, but it has also been recorded from coastal mangrove swamps and cactus forests. In Puerto Rico it is a rare bird, suffering, as other West Indies Crows, from the cutting down of forest trees. The White-necked Crows of the two islands are structurally indistinguishable.

This Crow bears a resemblance to the White-necked Raven of southwestern North America. Its contour feathers and those on the throat, upper flanks, breast, nape (where they are most conspicuous), back and abdomen have white bases, and the under parts are violet and purple. It is smaller than the White-necked Raven with a smaller bill and wings. In the White-necked Crow the nostrils are partly or completely exposed and its iris varies from light reddish brown to bright orange red. The White-necked Raven occupies a very different habitat of arid deserts.

In voice this Crow resembles the Ravens; it has a raucous, uncrowlike series of notes, its common call being a throaty 'culik-calow-calow'; it also gives a high-pitched 'kloak' or deep 'wallough'.

In Australia, where *Corvus* is the only genus of the *Corvidae*, five species are recognised, two of which are Crows and the other three Ravens. All five have the distinctive characteristic of possessing white eyes when adult, which suggests that the Australian species stem from similar genetic stock and may be regarded as a species group.

Crows occur throughout Australia from the arid interior to the

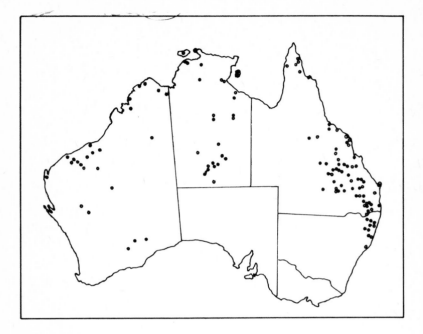

39 The Australian Crow
    ● The location of breeding specimens and nests of the Australian Crow
      Corvus orru
    ○ The location of non-breeding specimens of the Australian Crow *Corvus*
      orru

summit of Mount Koscuisko in New South Wales, and from Cape
York to southern Tasmania. The two species are the Little Crow,
Corvus bennetti, and the Australian Crow, *Corvus orru*, of which *C.
orru cecilae* is the only race in Australia. The other forms, *C. orru
orru*, *C. orru insularis* and *C. orru latirostris* occupy respectively
Moluccas, New Guinea and its southern islands, New Britain, New
Ireland and the surrounding islands, and Tannibar and Barbar
islands.

    Australian Crows are distinguished from Ravens by their smaller
size and the snowy white bases to their feathers in comparison with
the dirty grey bases of the Ravens. The colour intergradation is
abrupt in the Crows but gradual in the Ravens. The Crow's skin
underneath the two branches of the lower mandible is well feathered
but in the juveniles of the Australian Raven, *C. coronoides*, it is bare.

The Raven's throat hackles are more developed than the hackles of the Crows which are no longer than 1½″ (3.81cm).

The Australian Crow, *Corvus orru cecilae*, has a longer wing and shorter bill and legs than *C. orru orru*. It is a smaller bird, weighing 1.25lb (566.3g) in comparison with 1.4lb (644.3g) in *C.o. orru*. There is not sufficient difference in Australia between central, northwestern, northeastern, and southeastern birds to justify creating different races, but central and western Australian Crows tend to be lighter and smaller in all dimensions except that they have larger bills. Should future specimens justify the separation of an eastern and western race these would become *C.o. queenslandicus* and *C.o. cecilae* respectively.

The Australian Crow's plumage is all black with a purplish gloss on the upper parts, except for a slight greenish gloss on the crown. The glossed green under parts merge into black on the abdomen and under-tail coverts. Except for the black central rectrices the tail and the primaries are glossed greenish with the outer secondaries purplish and the inner greenish. The third, fourth and fifth primaries are strongly emarginated and the tips of the tail feathers are rounded. The Crow's eye is white with a blue inner ring and it has black feet and mouth. The female is smaller in all dimensions.

The habits and ecology of the central and western Crow population differ from the northern and eastern. The Crows of the arid centre and west are confined largely to isolated ranges and major watercourses where large Eucalyptus trees flourish under an effective rainfall. The Crows nest in these trees and tend to be residential, unlike their smaller relative, the Little Crow, which co-exists with the Australian Crow throughout the southern two-thirds of its range. In the better watered country of the north and east the Australian Crow is more common, and large flocks are frequent, especially in cultivated areas. Here the eastern population scarcely overlaps with the Little Crow.

The Australian Crow breeds in October, later and nearer the equator than the Raven. The nest is therefore necessarily well hidden under foliage for shade. It is the lightest of the Australian Corvid's nests, built of slender twigs and scantily lined. The Crows rebuild old nests. The nestlings' mortality may be reduced under tropical conditions because starvation of the surplus nestling only causes it to grow more slowly. It does not suffer from chilling which occurs in temperate regions. The eggs of the Australian Corvids are bluish or greenish, speckled with a darker colour and the Australian Crow's eggs are smaller in the east (42.6 × 29.4mm), than in the western and central regions (45.5 × 29.4mm). The juvenile's under parts are blackish-brown and have a soft loose texture; it has a grey iris, and the pink gape, mouth and tongue gradually become black. The second-year bird has a hazel iris and black soft parts. The immature Crow, unlike its parents, is nomadic.

Breeding Corvids in Australia are parasitised by the Australian cuckoo, *Scythrops novaehollandiae*; the young parasite is often raised together with the host's young although it is grey like the adult cuckoo and not black.

Unlike the majority of Crows this Australian species has no 'a' sound in its vocabulary; it gives a nasal, high-pitched, clipped, 'uk-uk-uk-uk', or 'ok-ok-ok-ok'. It has a distinctive habit of rapidly shuffling its wings above its back before settling them along its flanks.

The Little Crow, *C. bennetti*, is smaller than the preceding species. It has similar plumage except that the tail is glossed blue purple with the two outer rectrices glossed green, and the wings more blue-purple than greenish.

This species is nomadic and is capable of breeding wherever conditions are suitable and the rainfall adequate. In the Northern Territory it has been known to breed as far north as Tennant Creek and Brunette Downs. In the centre little is known about its distribution; it is common within a radius of 200 miles (320km) of Alice Springs. It has been recorded breeding eastward near the Diamantina River, Queensland, and Cobar and Ivanhoe, New South Wales. The recent clearing for agricultural settlement in the south and southeast has favoured the Little Raven, *Corvus mellori*, to the detriment of the Little Crow. At present it is rarely seen south of the River Murray in New South Wales, and it seems to have been replaced by the Little Raven throughout the South Australian wheatlands. In the southwest where the Little Raven is absent, the Little Crow breeds as far southwest as the 20in isohyet (rainfall measure), south of Shark Bay and along the River Murchison.

The substantial nests of the Little Crow are built in loose colonies

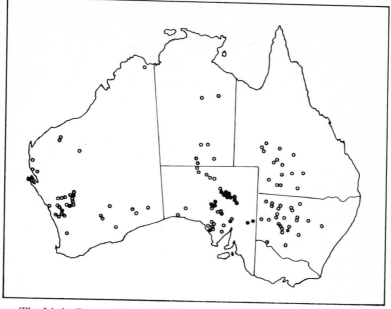

40 The Little Crow or Bennett's Crow
● The location of breeding specimens and nests of the Little Crow or Bennett's Crow *Corvus bennetti*
○ The location of non-breeding specimens of the Little Crow or Bennett's Crow *Corvus bennetti*

at the junction of three branches in tall shrubs and low trees and even on telephone poles; old nests are used the next season. Sometimes the Australian Raven breeds in the same vicinity without conflict. In areas where the Little Crow and the Little Raven converge they tend to develop separate colonies. This Crow's eggs are smaller but otherwise similar to those of the Australian Crow. The immature bird has dull purple upper parts and its tail and wings are glossed green. Its iris is blue-grey. First year birds are similar to the Australian Crow with sooty brown under parts; in the second year the only difference is that the tips of the primaries, secondaries and rectrices are brownish-black.

Sometimes mixed flocks of the Australian and Little Raven and the Little Crow occur together in the non-breeding season, but usually there are only two species. The Australian Crow and the Little Crow have been observed in the same region but the more solitary Australian Crow is outnumbered by the smaller species.

The Little Crow has a very nasal call: 'nark-nark-nark-nark', each syllable longer than that of the Australian Crow, and with the 'a' sound which is missing from the latter. No clear-cut morphological character has been found to separate the Australian Crow from the Little Crow, but ornithologists are convinced that their ecology, behaviour and calls justify their separation into two species rather than two races of one species.

Chapter 8

RAVENS

The Raven is the largest of the *Corvidae* family, although it is exceeded in length by the Magpie Jay of Central America. Its usually all-black form is distinguishable from the Crow's by the slightly larger body and a more rounded, graduated tail; also by the more pointed, narrower wings, and pointed, hackle-like throat feathers, which, in common with the feathers over the dark brown eyes, can be raised at will. Its massive black bill is curved at the top with bristles covering the nostrils.

The Raven hops and walks sedately, but gives cautious, sideways jumps when approaching a carcass, with its wings half-spread, ready for immediate take-off. Its voice is distinguishable from the Crow's hoarse 'kraak', by its guttural, croaking 'pruk-pruk-pruk' as it flies. It also has a deep 'Corronk' and a triple high-pitched 'toc-toc-toc'.

Unlike the Crow the Raven has not adjusted well to civilisation and is found only in the wilder, uninhabited parts of its former range. Once it was widespread in Europe, northern Africa, Asia (except the south and southeast), Iceland, Greenland, Australia and North America. Its natural habitat is mountainous, hilly districts and wooded lowlands, which it still occupies in Europe and is beginning to inhabit in British lowlands. It is now found in such dissimilar terrain as the waterless Sahara; the coniferous forests of Canada and Siberia; the seacliffs of North America and Scandinavia and the tundra islands of the Arctic Sea. It ranges from sea level to mountain tops, and seems to live anywhere up to 14,700ft (4,480m).

The Raven's direct and powerful flight has regular wing-beats. At 30mph, it is slightly slower than the Crow but it reaches greater heights, and glides and soars. In common with the Crow it indulges in aerial acrobatics. Such antics as nose-diving with closed wings, somersaulting, turning, twisting and tumbling, are used in play as well as in courtship. The male also glides upside down for a short distance in a cork-screw roll. A Raven has been seen lying on its back with a pebble clasped between its black feet, having held it thus when erect and then slowly toppling over.

Play forms a large part in its behaviour; Ravens pull each other's tails or play with each other's beaks as though 'kissing'. They pass stones to each other, sometimes a number of birds passing them from beak to beak, pounding and grabbing at them; or they offer a stick and then jerk it away in a teasing manner. The Raven is said by an authority to have a greater fund of conscious humour than any other bird. One Raven was seen baiting a dog; as the dog rushed for it it flew a little way, then settled; this continued until the dog was exhausted. Frances Pitt, a naturalist, owned a pair of Ravens named Ben and Joe. They had their own method of dealing with cats. Ben would parade close to the cat, who, fascinated by visions of a tasty meal, watched him closely. Joe would come up behind the cat and peck its tail. As the cat turned to see who was attacking it, Joe would strut off, and Ben would seize the cat's tail. This would continue for some time with the cat spinning round like a top.

The Raven's high intelligence is well known. It is easily tamed, a good mimic and it can count. Under suitable conditions it can live up to fifty years and a Common Raven has been known to reach sixty-nine.

A Raven in a wire-netting cage was freed by wild Ravens; they dug a hole under the wire and the caged bird dug from inside until the hole was big enough for him to escape. Another Raven with a damaged wing was nursed by an inn-keeper. Its cage was next to the hens, and it dug a hole under some rotting skirting board, and also pecked out a smaller hole and baited it. The hen came to the bait, and the Raven, through the larger hole, seized its beak and tried to drag it into his cage, in the process injuring it beyond recovery.

Caution is inherent in Ravens; most mature Ravens will approach a trap only when some creature is in it and they can see no danger exists for them. In the Shetlands 800 Ravens were attracted to stranded whales. A Raven has been seen to chase a rock-pigeon and

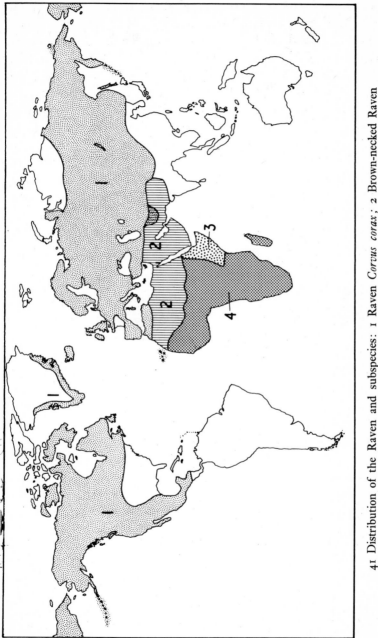

41 Distribution of the Raven and subspecies: 1 Raven *Corvus corax*; 2 Brown-necked Raven *Corvus r. ruficollis*; 3 Dwarf Raven *Corvus r. edithae*; 4 Pied Crow *Corvus albus*

catch it in mid-air. The two birds tumbled to the ground and the
Raven tore open the pigeon's throat. In common with Crows,
Ravens steal golf balls and deposit them elsewhere, probably
mistaking them for eggs.

Mated Ravens hold a territory and their mere physical presence
is enough to stake a claim. They mob flying predators either singly
or in pairs, flying repeatedly at passing Peregrines or swooping over
a heron on the ground, giving their hard, grating threat call of
either a cry or harsh, rattling note. They have a reputation for being
cowardly but they will show aggression to predators. The cock
Raven is master of the air on the sea coast. He will roll over crying:
'Krok-krok-krok', among a group of gulls, then suddenly utter
'KONK' and cock his beak at the sky. This may be a warning that a
Peregrine falcon is diving at over 120mph (200kph) and all the
birds will scatter except the Raven who flicks over on his back and
holds open his beak in defiance as the falcon passes him.

Corvus corax corax breeds in Europe. In the British Isles it is
now mainly confined to the west and north, mainly in hilly coastal
districts. Its range is from the Isle of Wight to Cornwall; thence
along the Devon and Somerset coast, to Gloucestershire and inland.
It has increased in Wales and is the hardiest of mountain birds in
Snowdonia; three pairs even nested in February as usual during the
Arctic winters of 1947 and 1963. Ravens are fairly plentiful over the
Welsh border to Hereford, Salop, Cumbria, the Pennine Hills and
the Isle of Man, but they are rare in the midlands and the eastern
counties. They have nested high up in the Douglas fir trees in the
Forest of Dean for twenty years, and are familiar on Lundy Island.
In Scotland they are numerous in the west, the Hebrides and
Shetlands and breed in smaller numbers in the north and south. In
Ireland they are resident mostly on the west cliffs and mountains.
This species also extends to the Caucasus, north and northwestern
Iran, Siberia, and is accidental in Spitzbergen and Novaya Zemlya.

The Common Raven has a glossy black plumage, iridescent in a
good light, with its crown glossed greenish or purplish and its back,
wings and tail giving a blue, green or purple sheen. The under
parts gradually change from a greenish gloss on the chin, to a
reddish purple on the upper breast and a green or blue purple on
the lower breast. The feather-bases are brownish-white like the

42 The Common Raven *Corvus corax corax* lives and breeds in the hillier coastal areas of Britain (*Eric Hosking*)

Rook's. It has numerous black, strong rictal bristles. The fourth primary of the pointed wings is the longest with the third no more than 3–12mm shorter; the first is between the seventh and eighth in length, and the second and fifth have emarginated outer webs. Except for the female being slightly smaller the two sexes are alike.

Breeding usually starts as late as the third year in Ravens. The mating display takes place early in the year and is quite an elaborate affair with the male raising the tufts of feathers above and behind his eyes and giving nasal muffled notes and cracking sounds with his bill. He then stands by the female and stretches his neck over hers, ruffling his head, neck and throat feathers he bows to her, sinking down on his belly, with his neck straight out but his bill pointing downwards; this is accompanied by curious noises such as mumblings, bubblings and poppings like corks being extracted from bottles. The pair then indulge in mutual preening. Stimulated, the male jumps in the air several times and caresses his mate's beak with his, tickling her under the chin and then 'kissing' her. Both birds stand like this until the female jumps up and down, a signal of her desire to mate. After copulation the pair celebrate with aerobatics and then land with the male giving a musical, vibrating call. As with other Corvids the pair mate for life and stay together.

Whether the nest is an old or new one both birds help to make it solid with sticks and heather stalks mixed and lined with mud and moss. Within the inner cup they place hair, wool and grass. The nests are usually built among high rocks or cliff ledges or in high trees, and the birds are solitary nesters. Some nests added to each year grow into stacks six feet high.

Breeding begins early in February or March often before the spring equinox. Ravens are almost unique among European birds in showing a decrease in clutch size from south to north because of the early breeding and a lack of longer days for feeding the young. The female lays daily one of the 4–6 blue to greenish eggs blotched brown or black, and, according to captive pairs studied, she starts incubating on the last egg and broods about twenty days. The male is in attendance throughout this time, greeting her with croaking 'corronks', and weak chattering 'pick, pick pick', before he feeds her. She replies with a querulous, high begging call like the nestlings.

The hatched young have fairly plentiful mouse-brown down above their dove-grey eyes, at the back of the head, on the wings and along the spine and thigh bones. They open their beaks continuously to reveal a purplish-pink interior and yellowish-flesh gape. They are fed by both parents and remain in the nest for 6–6½ weeks, the longest fledgling period of the British *Corvidae*. The Raven's extra long neck allows its throat pouch to distend so that it can carry food long distances. Its conspicuous, baggy throat when filled with food for its young is accentuated by its shaggy, pointed feathers.

After the initial brooding of the young the parents roost away from the nest and this has given rise to the idea that Ravens are unkind parents, but it must be remembered they are large nestlings and remain in the nest for a long period. In the Book of Psalms reference is made to the Raven nestlings: 'He giveth to the beast his food and to the young ravens which cry.' A cruel mother in Germany is known as a *Rabenmutter*: and in literature and tradition

Ravens are alleged to neglect their young, giving rise to the collective noun for Ravens being: 'An unkindness of Ravens'.

Few predators venture near a Raven's nest when the parents are in attendance; when the opportunity arises Jackdaws and Herring Gulls are the main egg-stealers. In Snowdonia the Peregrine falcon and the Raven are combative during the nesting season; the Peregrine harasses the Raven to protect its own nest, and often uses a Raven's old nest to breed in.

The fledgling's dove-grey iris turns dark brown some months after it has left the nest. Juveniles have duller plumage than adults, but otherwise they are similar except for their upper parts and lower parts being brown-black and brown respectively with some greenish gloss on the chin. During the first winter the tail, wings and wing-coverts remain browner. Soon after becoming independent from their parents the young disperse to find suitable habitats, forming communal roosts and joining non-breeding adults during the first autumn and winter. These flocks persist in a lesser degree during the breeding season in areas where there are no settled mating pairs. Over one hundred birds in a roost is quite normal, and a flock in the Hebrides contained over 1,200 birds. In Scotland there is a regular autumn and spring migration and long movements of up to 190 miles (304km) have been recorded between Scotland and Wales. Small flocks also migrate to Ireland. Apart from these movements Ravens do not move around a great deal and during the first winter young rarely move more than 20 miles (32km). The birds disperse during the day and gather in roosts in the evening.

The parents start to moult when the young are hatched and continue for 140–150 days, ending in September. Towards the middle of the period, when the primaries are moulted, the body plumage is moulted and renewed. In common with other passerine birds the wing and tail feathers moult when the fourth and fifth primaries have been dropped, and they finish dropping after the primaries.

Anything edible in forests, fields, shores or deserts will attract Ravens. They depend largely on the lambing season, devouring the placentas of ewes, carrion, and sickly lambs. They will flock around a large carcass like vultures, stripping its flesh and thus depriving blowflies of breeding grounds for maggots which harm sheep. Some mammals, and all small living creatures are preyed on; eggs, dead fish, insects and molluscs, corn, acorns and seeds all form part of their diet. They hold their food with their feet when eating, and bury any surplus, covering the holes with earth and grass.

There appears to be no recent decrease in Raven numbers through their general dependence for food on the carcasses of sheep dipped in organochlorine insecticides and they may be less sensitive to these chemicals than birds of prey. A digestive enzyme in the Raven's stomach has a similar action to the mammalian pepsin which breaks down proteins into proteosis and peptones. Hydrochloric acid secreted by the acid-producing cells of the digestive glands penetrates the food and prevents any further conversion of starch into sugar by the diastatic enzyme and also gives protection against harmful bacteria.

Persecution of Ravens by farmers and gamekeepers has been so great that in wide areas of central, western Europe and England the birds have disappeared from their old haunts. In 1822 some churchwarden's accounts at Talyllyn church in Dys-ynni valley, Wales, was recorded: 'Humphrey Jones for killing 3 Ravens . . . 2/-'. In 1913 a dozen Ravens together was recorded as an extraordinary gathering in the British Isles, and by 1914 they were uncommon in lowland Wales as they had taken to the mountains where they were safer. The pressure of persecution was lifted by the 1914–18 war and in recent years flocks of 40–50 birds have been quite normal. Although they are protected by law in Britain they are regarded as a threat to livestock and game and still shot, although they do not normally attack healthy sheep and lambs. The adult mortality is low because they are superior in strength and size to all perching birds and as they start to breed late they are fully matured and experienced. The young die by flying into overhead cables and road traffic; the mated pairs cannot avoid taking greater risks during the breeding season and some die from territorial strife and the strain of breeding. The disappearance of these strong, handsome birds would be a tragedy and they deserve to have their legal protection honoured.

The allied forms of *Corvus corax* have variations in form and bill. *C.c. hispanus* of south-east Spain has a shorter wing and a generally more arched bill than the British Raven; and *C.c. varius* of Iceland and the Faroes has a larger wing and thicker bill. *C.c. tingitanus* of North Africa and the Canaries has a shorter, stouter bill and less elongated throat hackles. The Raven in eastern Greece is probably closer to the Punjab Raven, *C.c. subcorax*. In the region between Lake Baikal and Yakutsk the Raven population is intermediate between *C.c. corax* and *C.c. kamtschaticus*, the Siberian Raven, which occupies the northeast of Russia southwards to Hokkaido.

The Punjab Raven, an enlarged, heavy replica of the Jungle Crow, is to be found from eastern Greece, eastwards to northern and western Iraq, Iran, Baluchistan, Afghanistan, northern India, Russia and Chinese Turkestan, where it replaces the Tibet Raven in the plains or lower mountains. Its shorter and more sharply pointed throat hackles distinguish it from the Tibet Raven, which has similar steely-blue black plumage and a massive bill.

The Punjab Raven waddles or hops clumsily in the vicinity of towns and villages as well as outlying hamlets and nomadic encampments, where it scrounges for scraps close to the occupants. It is quite tame in out of the way desert habitations and along caravan routes. Pairs usually hunt together but fair numbers will collect where food is plentiful, often joined by kites and Egyptian vultures, to feed on carrion. It shares communal roosts with kites and House Crows, flying in twos and threes from all directions to congregate in considerable numbers at sunset.

In the air its strong, direct flight is made distinctive by the wedge-shaped outline of its tail and the loud and peculiar creaking of its wing-quills. Occasionally pairs or small parties take off on rising air currents to circle on motionless wings or to perform remarkable aerobatics.

This Raven's common call is a hoarse, wooden, bell-like 'caw' and 'pruk pruk'. It has a large vocabulary of vocal expressions, some quite pleasing and musical.

Nests are re-used year after year on the same site near the tops of trees such as Dalbergia and Acacia. The Raven is a solitary nester and will breed in semi-deserts in stunted trees, or on a railway signal tower if no other site is available. The incubation period of the 4–6 eggs (50.7 × 33.6mm), is 17–18 days with both parents taking part in the breeding. The first year birds have less gloss and browner wings and tails than their parents. These Ravens are very long-lived and, in captivity, one has lived to be over seventy.

The largest of the Indian Ravens is the Tibet Raven, *C.c. tibetanus* which is distinguished by its massive bill and strongly lanceolate throat hackles. High altitudes are the habitat of this trans-Himalayan species; it was recorded by the first Mount Everest Expedition around their Camp 3 at 20,997ft (6,400m), and it normally occurs between 13,123ft (4,000m) and 15,404ft (5,000m) in the Tibetan plateau. Its range covers the mountains of central Asia from Kansu to Tibet and the Tibetan tracts of the Himalayas, also north to the Kun Lun Shan in northern Tibet.

In flight this Raven's wing quills look slate-coloured, contrasting with the jet-black under-wing coverts. It revels in spectacular aerobatics, gambolling in high winds, gliding in formation at speed, wing-tip to wing-tip, or one behind the other, looping the loop, nose-diving deep into the valleys, then zooming effortlessly up again.

The wild, wary and suspicious Tibet Raven becomes bold when scavenging around villages or searching for food in nomadic traders' encampments. It joins Griffon and bearded vultures to feast on carcasses of pack animals along the caravan routes. It has been seen to swoop on a full-grown Tibetan partridge and bear it away. Its call is similar to that of the Punjab Raven, a wooden, bell-like 'prak-prak' or a high-pitched, guttural 'kreeuk'.

Often when the country is still covered in snow this Raven will start breeding, as early as February or March, at an altitude of 3,300 to 4,500 metres, with its nest placed near the top of an inaccessible cliff. The 3–6 glossless, pale bluish-green eggs, (Punjab: 47.5 × 33.1mm; Tibet 51.75 × 35.5mm), have grey smudges and blackish streaks. They are laid in a thickly lined cup of wool, yak and other mammals' hair.

When the Punjab Raven's plumage becomes brownish with wear it is not easy to distinguish it from the visiting Brown-necked Raven, *Corvus ruficollis* of the Sahara. The latter is smaller and has a brown, coppery tinge over its head, neck and undersides, with the rest of its upper parts, wings and tail a glossy violet black. Its bill is usually more slender, with the nasal plumes at its base shorter than those of the Punjab Raven. Some Asian populations of the *C. corax* species are similar to the Brown-necked Raven in being smaller and browner, and these two birds were long regarded as conspecific; but in Palestine and North Africa they are ecologically segregated with the *C. corax* species occupying the hills and more humid country, and the Brown-necked Raven in the pure desert regions scavenging round nomadic encampments. The Brown-necked species has a wide range stretching from Cape Verde Archipelago and the whole of the desert and semi-desert regions of North Africa further east to Iran, Afghanistan, Sind and Baluchistan.

This shy bird becomes tame around villages and encampments. It is found in large numbers at Khartoum and parts of Kordofan and the Red Sea coast. Its call is a loud 'cronk'. It appears to nest as far east as the Sind–Baluchistan frontier hills from January to March. In western Sudan it breeds from December to March, and in Manchuria it breeds in October. It builds in trees or on rocks and in Baluchistan nests have also been found on ledges of clay cliffs. The 3–6 eggs (45 × 30.9mm) are similar to those of most Ravens and are laid in a nest lined with camel hair and soft materials as available. The juvenile is less bronze than the adult with its head and under parts dullish black.

The Brown-necked Raven and the smaller Dwarf Raven, *C. ruficollis edithae*, are the direct link between the superspecies, the White-necked Raven, *Corvus albicollis*, and the Pied Crow, *Corvus albus*. The Dwarf Raven of Ethiopia, Somaliland and Kenya, is isolated from the Brown-necked Raven and is a much smaller bird with a relatively shorter bill and white bases to the nape feathers. In the latter feature it resembles the Pied Crow, a bird of similar size, which bears a white hind-collar and a broad white band across the breast; but the Pied Crow's bill is relatively longer like that of the Brown-necked Raven. The Dwarf and White-necked Ravens have been known to hybridise in Eritrea and southern Ethiopia, but the species are mostly segregated with the White-necked Raven confined to areas of human habitation along the edges of the desert; it also penetrates the range of the Brown-necked race at Saharan oases.

The Dwarf Raven is more like a Rook than a Raven, and 'caws' like a Rook. It is an inveterate scrounger, invading huts and kitchens

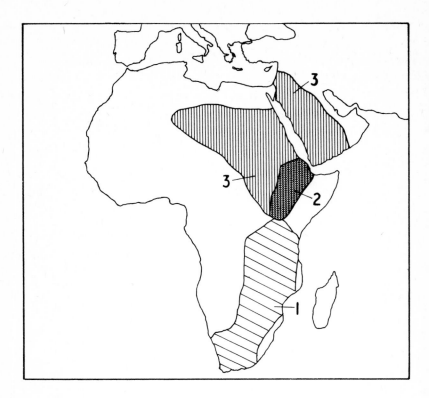

43 Distribution of the White-necked Raven and subspecies
  1 White-necked Raven or Cape Raven *Corvus albicollis*
  2 Thick-billed Raven *Corvus crassirostris*
  3 Fan-tailed Raven *Corvus rhipidurus*

for food and indulging in bathing whenever it discovers water in its arid regions. It is a loose colonial nester unlike the larger species, and builds a typical Crow's nest in tall saplings or on cliffs and surrounding bushes. Its 3–5 eggs are pale blue, spotted dark brown (42 × 30mm), and they are laid in February in northern Kenya and from April to June in Somaliland. Because of its colonial nesting and distinct call some ornithologists think the Dwarf Raven should be recognised as a full species.

The African species of Raven are more varied than those of other continents and all but one show strong affinities with the Common Raven. The White-necked Raven, together with the Thick-billed Raven, *Corvus crassirostris*, and the Fan-tailed Raven, *Corvus rhipidurus*, are all thick-billed birds. The first two are mountain-

44 The Thick-billed Raven *Corvus crassirostris* occurs in northeast Africa
(*Zoological Society of London*)

birds, but the Fan-tailed Raven inhabits lower altitudes over more
arid country and there is a similar variation in the *Corvus corax*
species. The Fan-tailed Raven's short tail has peculiar fan-tailed
uropigial (rump) bristles; the White-necked Raven's tail is relatively
short and bridges the difference between the Fan-tailed and the
Thick-billed Raven whose tail is longer.

The true representative of the Raven in Africa is the Pied Crow,
Corvus albus. It is also thought to be the African representative of
one of the Asiatic groups of *Corvus* and therefore should be placed
near *Corvus torquatus*, which spreads from eastern China as far
north as Hopeh to North Vietnam and Hainan Island, and westward to eastern Sikang. The Pied Crow is a distinctive species,
slightly larger than the Carrion Crow, with a bill as massive as a
Raven's.

The distribution of the Pied Crows is wide, but they are mostly
local and very common where they occur. Their range stretches
from Senegal to Sudan southward. In West Africa they are to be
found on the mainland except in true forest or treeless desert. They
inhabit the islands of Fernando Po, Zanzibar, Aldabra, Assumption,

Comoro and Madagascar. They depend on human refuse so populated areas are their favourite haunts. Their wary, but impertinent personalities bring them into close contact with man. They will unearth and eat seeds as a planter plants them and walk through ripening corn until they spot a desirable cob, then jump on the stalk, break it, stand on the cob and skilfully peel it to extract the corn. They snatch farmyard chicks; destroy thatched roofs by scratching around them for grubs; raid nests of palm swifts, weavers and waxbills, and help themselves to the farmer's cassava and palm-oil nuts. They can, however, be beneficial to man, cleaning up the rubbish dumps and the carrion, eating insects such as wireworms, cockchafers, the larvae of fly, beetle and moth; termites, ticks, locusts, caterpillars, snails and even small tortoises, frogs and mice; also dead fish, crustacea and molluscs. In Liberia they in turn are eaten by the natives.

Rain plays a large part in the Pied Crows' habits. Breeding is triggered off by the onset of rain so that their nesting period is variable. They start breeding as early as March in the northern Congo where there are tropical rain forests; in Senegal and the northern regions they breed from May to July. In the more southern parts of their range breeding is mainly from August to September and even as late as October and November in Tanzania and Rhodesia. Before the rains they indulge in aerobatic displays soaring and mobbing. They nest mostly in palms 70–100ft (21.3–30.4m) from the ground, also in baobabs, silk cotton trees and small thorns, and are almost unique in nesting in the introduced eucalyptus trees.

The breeding pair line their 18″ (46cm) diameter platform of stick and roots with wool, hair, string, rags and large feathers to make the 3″ (7.62cm) deep cup of about 1″ (2.5cm) thickness. The 3–6 bluish-green eggs, spotted brownish-grey (44 × 30mm) are incubated soon after the first one is laid. Like the Indian House Crow, the Pied Crow is parasitised by a cuckoo – the Great Spotted Cuckoo, *Clamator glandarius*, which may lay as many as three eggs in the host's nest. The adult cuckoo is unlike its host; it has a pale grey chest, brownish-grey upper parts spotted with white, and is smaller than the Pied Crow. In its attempt to mimic the host egg the cuckoo lays eggs that are 7% of its body weight, and although

they are 25% smaller than the Pied Crow's eggs they are particularly large for the cuckoo. The young parasitic birds also mimic
the host's young by being black on the upper part of the head and
nape with grey plumage that is darker and less spotted than the
adults'; their primaries are chestnut. The host's young are too big
for the cuckoo nestlings to eject. The Crow's fledglings leave the
nest after five weeks. The immature Crows have brown freckling
on the white of their plumage and dull black from the chin to the
breast and under parts.

The erratic migratory habits of the Pied Crow are governed by
the rain. After breeding they move from the rainy areas to drier
ones and wander outside their normal range to the southern
borders of Libya or the northern Somaliland coast. They have
spectacular morning and evening flights and gather in flocks of a
hundred to circle high in the air to avoid being distressed by the
midday heat. At a temperature of 37–39°C (98–102°F) they keep
their beaks open when flying at a low altitude. At night they roost
communally.

This species is a good mimic and learns to talk readily in captivity.
In the wild it gives a hoarse, guttural 'kwaww', almost like a snore,
besides growling and guttural noises and toad-like croaks changing
to a tinhorn note 'k-k-k-ko-o-a-a-ah-h-h'.

The close relationship between the superspecies, the White-
necked Raven, and the Thick-billed Raven is apparent in their
heavy, relatively short, white-tipped bills. Both species have
shorter and less forward-growing bristles and shorter feathers on
the hind-neck than the Common Raven and the throat hackles are
bifurcate and not lanceolate. They are often regarded as a separate
genus, *Corvultur*, to distinguish them from *Corvus*. The Thick-
billed Raven has less white on the nape and more glossy plumage.

On the large and formidable rocks and cliffs of the southern and
eastern regions of Africa the White-necked Raven is widespread.
Pairs or large flocks travel long distances in search of food and
perform wonderful aerobatics, their wings making a great noise,
similar to that of the Punjab Raven. In South Africa and Angola it
is known as the Cape Raven, to differentiate it from the White-
necked Raven, *C. cryptoleucus* of North America. In common with
the Thick-billed Raven it has a good sense of smell and is the first

to find dead animals, even leading the vultures to their feasts. This bold and daring species is well versed in self-preservation, but it soon becomes tame and fearless in built-up areas. The white collar on its hind neck and the few white feathers on the chest distinguish it from the other Ravens. It has a bronze-brown sheen on the head and foreneck but is otherwise black. Its deep, heavy, white-tipped bill is smaller than that of the Thick-billed Raven. Its curiously high-pitched tin-horn croak is familiar in the Ruwenzori mountains between Uganda and the eastern Congo, where it has a prolonged breeding season starting in February. It breeds later in the season further south. In Kenya the breeding season is from October to December and further south in Malawi from September to November.

The same large nest of sticks is used every year; it is built on a ledge or cliff and occasionally in trees. The 5–6 eggs are glossy blue-green spotted drab (52 × 34mm). The back of the sooty-black juvenile's neck is white streaked with black.

The Thick-billed Raven is a common sight perched on rocks and stumps in the highlands of Ethiopia at a height of 4,500ft (1,372m) in the Kivera Hills and in southern Ethiopia. It is also to be found at Eritrea and wanders to southern Somalia and the Sudan, appearing when camps are moving to pick up the scraps, and smelling out dead animals. It gathers in flocks and damages crops, but on balance it is looked upon as beneficial for ridding the land of carcasses. It is larger than the White-necked Raven with a longer, much heavier bill and longer tail. A white patch on its head is joined by a white streak to the white on the back of its neck. Its habits are similar to those of the White-necked Raven and it has the same call, except that the hoarseness of its croak sounds as though it has lost its voice and is suffering from a sore throat.

During the breeding season from December to February these Ravens indulge in grand aerobatics. They nest on rocks and trees. The juvenile's plumage is less glossy than the adult.

The Fan-tailed Raven is a more divergent member of the super-species; it has no visible white on the nape although the nape feathers are white-based. The bill is more like that of the *C. corax* superspecies in colour and shape, having no white at the tip, but it diverts in being short in relation to the Raven's size. Its nasal bristles and head feathering are similar to the other two African species. It is the smallest of the three Ravens and has a wholly glossy, black plumage with a purple and bronze sheen. This Raven is not confined to Africa, but spreads from the mountainous areas of the south Sahara into the Sudan, Ethiopia, Somalia, Palestine and Arabia. It is found in Yavello, southern Ethiopia and outlying districts, where it goes in search of offal. It lives on rocks and cliffs and travels far for food. It is identified by its short tail as it flies in pairs or soars into the air in flocks. When walking this Raven has a curious habit of opening its bill as though panting. Its cheery falsetto croak is often heard near built-up areas where it scrounges around camps and villages for food. It also has a lower note which comes deep down from the stomach and can only be heard close at hand.

The nests of Fan-tailed Ravens are built in inaccessible holes and crevices of cliffs; the birds are colonial nesters and breed in the Sudan in May and June. Their 3–4 eggs are similar to those of the Brown-necked Raven but slightly larger (45 × 32mm).

There are three species of Raven in North America. The Common or Northern Raven, *Corvus corax principalis*, is slightly larger than the Eurasian species. It resembles the Common Crow but it is larger with a much heavier bill and the feathers at the throat are elongated and pointed. When it takes off this Raven hops two or three times in the air, whereas the Crow jumps directly upwards. In flight the Raven's longer neck and bill extend farther ahead of its wings and its tail is rounded or wedge-shaped. It glides and soars more than the Crow and does barrel-rolls, dives and tumbles. Its voice is very different from the Crow's; it has a hoarse, far-carrying, rather wooden 'kwaawk' and other notes including a bell-like call.

This Raven favours mountainous and wild hill country and sea-coasts in the Arctic and forested regions. Its range stretches from the islands of the Bering Sea, the Aleutian Islands, Alaska, Arctic Canada, coastal Greenland, and central and southeastern British Columbia; south along the coast and through the centre of the States of Alberta, Saskatchewan, Manitoba, the northern parts of Minnesota, Wisconsin and Michigan to southern Ontario, central and northeastern Quebec, southeastern Maine, southern New

Brunswick, Nova Scotia and Newfoundland. It is also found deep in the Appalachian mountains as far south as northeastern Georgia. It formerly bred in the Great Plains and in northern Arkansas and northeastern Alabama.

In the Arctic this Raven's breeding distribution is governed by the presence of cliffs, and in Newfoundland and the Labrador coast it favours cliffs overlooking the sea. It often uses the same rocky nesting-places as seabirds. In woodland areas it nests solitarily, building a large nest in conifers and other trees. Breeding takes place as far south as the Allegheny mountains and the evergreen, coastal islands of Maine, where only one pair occupy each island.

The twenty-day incubation of the 3-5 eggs is shorter than the European species. The very noisy young have a dull, brownish black plumage with long, pointed throat hackles. Their voices are higher pitched than the adults' and they make more prolonged calls.

The Ravens' habitats do not greatly overlap with man but they prosper in such heavily farmed districts as Annapolis Valley and visit subarctic towns and villages, scavenging at garbage dumps. They are common in some sections of the North Carolina mountains, regularly visiting rural slaughter pens for food. When not persecuted they become tame and forage over a wide variety of terrain, favouring the lake shores, rivers and sea coasts. In Yellowstone and Glacier National Parks, along with the bears, they plunder the garbage bins at the back of hotels.

Careful observation has practically exonerated Ravens from the charge of attacking game birds' young and eggs; their scarcity prevents them from doing large-scale damage, but they rob Gulls' and Terns' nests. They are not regarded as destructive to crops. The mortality of the Alberta Ravens was high when strychnine was put down to kill the wolves, but this practice has now been stopped.

A smaller species, *Corvus corax sinuatus*, with a slightly larger leg than the Northern Raven, occupies mountains in western North America from the Okanagan lake valley in British Columbia through northern Idaho, south Montana and South Dakota to southern California; also through central America as far south as north-western Nicaragua. A subspecies, *Corvus corax clarionensis*, similar to *sinuatus*, is resident in Baja California, the Gulf Islands and Revilla Gigedo Islands.

The White-necked Raven, *Corvus cryptoleucus*, occurs in the hot southwestern deserts and valleys of the western United States and Mexico. This smaller species is often called a 'Crow' because of its crow-like habits. Its black plumage has a violet sheen and white bases to the feathers of the neck, lower throat, chest and breast. Its numerous nasal bristles point forward, covering the nostrils of its long bill; it has pointed throat feathers and stout legs and feet. Its long, pointed wings are longer than its tail.

The White-Raven's curiosity makes it an easy target for the shooting range but after its flock has been shot at it becomes timid and suspicious. Where it has not been molested it is quite tame and friendly. It has no fear of wagons and horseriders and visits schools to scavenge the remains of children's lunches. During the cold weather small flocks gather in stockyards to feed and they also invade cities where they are welcome as useful scavengers.

Breeding pairs build their crow-like nests in trees and bushes quite near the ground. The 4–7 pale bluish-green eggs are spotted brown with zig-zag lines of olive-grey. Even during the breeding season Ravens are often seen on the tableland and during winter they gather in flocks, feeding on animal matter and locusts.

Australia has three species of Raven. The largest, the Australian Raven *Corvus coronoides*, and the smallest, the Little Raven *Corvus mellori*, occur throughout most of southeast Australia, and the Forest Raven, *Corvus tasmanicus*, is found in Tasmania, Flanders Island and the Victoria mainland, and the New England Raven, *Corvus tasmanicus boreus*, occurs in New England mainland.

A great deal of research has gone into the Ravens' and Crows' status in Australia and they are regarded as being of economic importance. Ravens eat a wide variety of food in Australia, depending on the season and district; some birds are mainly residential but most move and feed in flocks, and it is these which affect agriculture and forestry. It has been found that Ravens feeding on crops in one season will be consuming hosts of injurious insects at another. This balance is evident in the lambing season; the Ravens' consumption of after-births and dead lambs clears disease-forming placenta and carrion. Ravens attack sickly and unattended lambs, eg when a mother bears twins and leaves one lamb alone, or when a mother gives birth for the first time and deserts her offspring.

It has also been found that farmers lambing outside the normal time may attract an abnormal number of Ravens, and as the available carrion does not satisfy them the Ravens turn to the lambs. Careless farmers who overfeed or underfeed the ewes, or who provide inadequate shelter or insufficient supervision during lambing, are partly responsible for Raven predation.

Over the past years Ravens have been controlled in Australia by shooting, blasting of roosts, scaring, trapping and poisoning, all of which methods have disadvantages. Poisoning by organic insecticides appeared to be the best method but it also kills dogs. Extensive research has shown that relatively few lambs are killed by Ravens and that the traditional destruction of these adaptable and useful birds would lead to greater outbreaks of fly-strike, thereby causing a greater loss of stock.

There is little else to distinguish the Australian Ravens from Crows other than their throat hackles and the dirty grey bases to their feathers. The ratio between the bill, tail and wing has been found a useful criterion for identification in other parts of the world but in Australia both Ravens and Crows include large and small species.

The Australian Raven is a large, shining black bird with a bluish sheen. Its outer primaries, primary coverts and alula are glossed green. The third, fourth and fifth primaries have strongly emarginated outer webs. Nasal bristles completely cover the nostrils. The prominent throat hackles are over 35mm long and reach the status of ornamental plumes; they are distinctly 'fanned' when the Raven is calling. There is also an extensive unfeathered area of black, loose skin under the chin. Its bill, legs and inside of the mouth are black, and its iris white with a blue inner ring. Except for the male being larger both sexes are similar.

In eastern Australia the distribution of the Raven corresponds with that of the sheep. The Raven appears to have been a vigorously expanding race which was spreading over southeast Australia before the advent of the European settlement, but sheep-farming may well have accelerated its population over the last 150 years. The short-grazed sheep pastures provide Ravens with easy access to insects and lizards besides carrion. There is a most conspicuous difference in the Raven's decrease in size from north to south in the

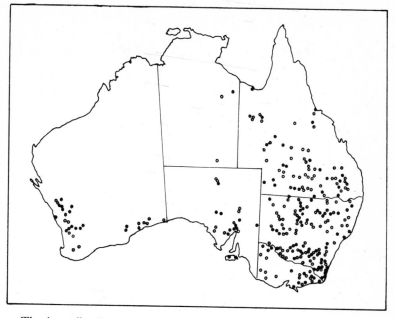

45 The Australian Raven
○ The location of breeding specimens and nests of the Australian Raven *Corvus coronoides*
● The location of non-breeding specimens of the Australian Raven *Corvus coronoides*

eastern side of the continent and also from east to west all over Australia. The greatest comparison is in the bill size, the largest-billed Raven being in Queensland and the smallest in West Australia.

It is not clear why the Raven in Western Australia does not follow the sheep-farming as extensively as its western counterpart. The aridity of Central Australia would not appear to deter these birds as they have a breeding population near Oodnadatta in northern South Australia under much drier conditions. Perhaps the western population is older stock and lacks the eastern race's colonising vigour. The western and eastern populations are joined by a narrow corridor north of the Great Australian Bight in the Eucla Basin.

Breeding Ravens occupy large territories of around one square mile to two pairs all the year round, but the non-breeders and immature birds live in nomadic flocks and travel long distances. Despite the wide range they occupy, breeding is confined to the spring, egg-laying being mostly in August. Courtship displays consist of aerial chases and mutual preening, but they are not so spectacular as those of the Northern Ravens.

When calling from its territorial perch the Raven fans its large throat hackles while stretching the head and neck forward level with its body. It swells its baggy 'gular pouch' area and this, combined with the fanned hackles, gives it a most characteristic profile. The gular pouch is in the unfeathered area under the chin and extends into the neck serving as an integral part of courtship display and as a resonating chamber in courtship. The very loud, territorial call is one of the best known bird calls in Australia and can be heard a mile away. It consists of four 'Cars', the last becoming a prolonged downward gargle. This distinguishes it from the Little Raven's call which lasts about the same time but has twice as many syllables and does not have the long drawn out ending. Other calls are alarm 'quarks', a panic or mobbing call, and a subdued chortle. Ravens also have a high-pitched wail like a child's cry.

Eucalyptus trees are the commonest nesting sites, their massive branches forming a shady canopy and a strong support for the large, bulky nests, weighing 10lb. The Raven pair chooses a site with a good vantage point 40–80ft (12–24m) from the ground, where it can watch for trespassers. The nest of twigs is lined with bark and wool about 1″ (2.54cm) in thickness, which ensures it is dense enough to resist charges from a shotgun.

The female is smaller than her mate with a smaller wing and tail ratio, and she has an extensive brood patch during nesting. She lays on average four pale-green eggs marked dark- and olive-green (46 × 30mm), which she incubates for 19–20 days. After hatching the young remain in the nest for 42–45 days. At fledgling stage they show a bare pink area under the chin but they lack throat hackles until they are 3–4 months old. The nestlings have brownish-black body feathers but soon moult them after leaving the nest. Their eyes are a gun-metal colour in the nest and brown soon after fledging. The juvenile is similar to the adult with a dull purple gloss and the under parts have loose, soft-textured feathers. The bill is black, tipped with brown, and the feet are blackish-brown; these change to black by the first year. The iris becomes firstly blue-grey, then changes to hazel before it reaches the adult white colouring at about 30 months.

During and after the breeding season the parents moult. The replacement of primary feathers takes seven months, each feather's full growth taking three to four weeks. The fledglings are very slow to mature; they leave their parents voluntarily six months after hatching and join flocks to forage through a district during midsummer. As a result of banding 11,558 Ravens during 1953–68 a very high mortality has been found among immature birds; 64% of nestlings die in their first year, mainly during the winter. They wander miles for food and come in contact with man who shoots, poisons and traps them. For two-and-a-half years they lead nomadic lives in flocks of about thirty birds; they feed on grasshoppers, pasture grubs, post-harvest grain and carrion during lambing. They are notorious for pecking out lambs' eyes; with one foot anchored on the dead lamb's head a Raven will lower its long and massive bill at right angles to its neck, then drive it into its target with force.

Second year Australian Ravens have notably less mortality than the Little Raven of the same age; but the latter matures in two years to the Australian Raven's three, which balances the discrepancy to some degree. Control of the Raven population largely affects the mobile flocks and most breeding pairs are unharmed and able to renew their losses. Some immature birds return to their original district and remain there all their lives. Breeding Ravens may live for 7–8 years.

In comparing the Northern and Australian species of *Corvus* it has been found that the Northern Raven, *Corvus corax*, and the Carrion Crow, *Corvus corone*, suffer mortality in their first year very similar to that of *Corvus coronoides*, ie 63.5%, 62.4% and 64.9% respectively. This is accounted for by the fact that they are all resident territorial species. The mortality of the Northern Rook and the Australian Little Raven is also similar – 54.0% and 55.8% respectively – and both suffer heavy mortality in their second year. These two species are ecologically similar; both are nomadic or migratory, colonial nesters, and feed in flocks.

It was not known until 1965 that there was more than one species of Raven in Australia. Much of the banding of Ravens was done before the Australian and Little Raven were separated as two distinct species. As the study of Ravens progressed it was found there were many ecological differences between the Ravens being studied at Toganmain Station in southern New South Wales and the Ravens on the Tableland near Canberra. Further study showed differences in call, territory, nomadism, behaviour, mating, measurements and plumage. It was also found that the smaller Raven occupied territory at Toganmain in common with the larger species which favoured the taller trees. To distinguish it by name from the larger Australian Raven, *Corvus coronoides*, it was decided to call the new species *Corvus mellori*. But the popular name is the Little Raven.

This smaller Raven resembles its larger relative in its black plumage and large, but less massive bill, with nasal bristles covering the nostrils. It has a well-feathered area under the chin unlike the larger race, and its hackles are smaller and usually bifurcate (forked), the feathers tending to separate rather than cohere. Its plumage has a bluish-purple sheen, with ear coverts greenish-black, the under chin glossed green, and lower hackles purple. The blue-purple tail has glossy green outer webs on the outer-most rectrices, and both Ravens have ashy-brown under-down and feather bases.

The Little Raven gives two or three rapid, vertical wing flips as it voices its very guttural 'kar-kar-kar-kar' or 'ark-ark-ark'. It has a more elaborate mating display than the larger species. It droops its half-closed wings, cocks its tail at an angle of over 45° and slowly promenades in front of its mate. Because it has a smaller breeding territory it has a ground display instead of aerial chases. Five pairs of Little Ravens have been found occupying two acres of land. They nest on trees and bushes, rarely above 30ft (9m), and often take up residence in dead trees or on telegraph poles. Both the nest and incubation period are similar to the Australian Raven, and except for being smaller, the eggs are similar (44.5 × 29.6mm). The female has the same extensive brood patch as the larger species but there is a shorter nestling period and the fledglings are quicker to develop.

Although the two Ravens' breeding areas are rather similar they do not infringe on each other's territory. The smaller species alone occupies the Southern Alps. It does not occur in West Australia, the Northern Territory, Queensland or Tasmania, but fluctuating numbers have been found on King Island, suggesting it migrates over 56 miles (90km) from the mainland. An isolated breeding population occurs above 5,000ft (1,524m) in the Kosciusko State Park, New South Wales, where snow gums, *Eucalyptus niphophila*, replace forest species, thus excluding the Australian Raven.

After breeding the family parties feed together and at dusk they join with other families to roost. Finally, the whole flock, often over one hundred birds, travel 200–300ft (61–91.4m) up, their purposeful flight keeping a straight course, to a fresh source of food in an adjoining roost. Although they occupy their breeding territory only for three months, they return to the same area every year. As with the Australian Raven, should the female die the male finds another mate and occupies the same territory, but the loss of the male partner causes the female to abandon the territory.

The juveniles resemble the adults but have brown-black feathers of a soft, loose texture on their under parts. The upper parts of their plumage are only glossed blue-purple on the scapulars, back, wings and tail. The first year birds are indistinguishable from the adults except for the brown eye and pink and black skin of their tongue, mouth and chin. Only traces of hazel in the iris and pink at the bases of the upper and lower bill and under the chin, mark the second year birds as not wholly mature. Over two years 61.3% of immature birds have been known to die annually.

Although the Little Raven does not attack lambs it competes unsuccessfully with the larger species for afterbirth and carrion. To counteract this it feeds on the ground consuming more insects and seeds. It has difficulty in removing dead lambs' eyes. During the banding study it was found that nearly half the Little Ravens were shot because they are less aware of human predators than the Australian Raven. Only one of the latter was found shot but the larger Ravens had succumbed to trapping and poison because of their attraction to carrion.

The species *Corvus tasmanicus* consists of the Forest Raven, *Corvus tasmanicus tasmanicus*, and the subspecies, the New England Raven, *Corvus tasmanicus boreus*.

The Forest Raven is found in Tasmania and the Bass Strait islands; and in southern Victoria in the Otway Ranges and Wilson's

Promontory. It has the identical plumage coloration of the other Ravens, except the inner webs of the primaries have their tips and trailing edges brown-black. The distinguishing features of this Raven are its massive bill and short tail which results in a very low wing-to-tail ratio. The inconspicuous throat hackles are not erectile, and the under-chin area has a small extent of bare skin visible. The Forest Raven's leg is larger than that of the other Australian Corvids, except for the Little Raven's. It is the heaviest Corvid in Australia, weighing 712 grams to the Australian Raven's 675.2 grams. The female is smaller than the male and has an extensive brood patch.

The juveniles and two-year-olds follow the same pattern in all features as the other Ravens but are slower to develop than the Little Raven and their eyes remain hazel into the second year with the tongue, mouth and chin still showing considerable pink coloration.

In Tasmania the Forest Raven ranges from the wet sclerophyllous (hard-leaved) forests to the savannah woodlands. It is a resident species and the only Corvid occupying this, and the adjoining islands. Two distinct populations occur on the Victorian mainland. The Wilson's Promontory population is separated from the rest of the population in the Otway Ranges by the heathland and plains of Yanakie peninsula, which is mostly treeless and unsuitable for breeding. In the Otway Ranges, 150 miles (241km) further west, the Forest Raven resides in the west sclerophyllous eucalyptus forests on the steeper hills and does not encroach on the northern open plains of the western district occupied by the Little Raven. It is thought the Forest Raven might be losing ground in Victoria against the vigorous spread of the Australian Raven.

The call of the Forest Raven is deeper than that of any of the Australian Corvids; its 'korr-korr-korr' is a slower, much deeper call than the Little Raven's, with a long drawn out last note. Where open paddocks occur in the Forest Raven's environment it forages for food and carrion.

The New England Tablelands, the only habitat of the New England Raven, *C.t. boreus*, is continually being cleared so that the area of wet sclerophyllous forest, which is its favourite haunt, is rapidly shrinking towards the eastern slope. The Australian Crow,

C. orru is the common breeding species to the east and north up to the foot of the slopes and its flocks fly westwards in the autumn. The two species appear to maintain their own territories during breeding. But to the west and south the Australian Raven meets the New England species along a wide front and is an aggressive competitor.

The plumage colouring of the New England Raven is similar to that of the nominate species; it also has the feathered chin, and inconspicuous, though rather rougher throat hackles. It differs from the Forest Raven in having a larger tail and wing, which is significant, particularly in the female, whose tail is 215.8mm compared with 198.7mm in the male, and whose wing at 373mm is longer than the male's 372.3mm wing; it is therefore as long as that of the Australian Raven. The male's leg is longer than the leg of any of the Australian Corvids. This Raven's large bill is not as massive as that of the Forest Raven. It is a heavy bird, with the female, in all, smaller, and the immature birds following the pattern of the Tasmanian Raven's sub-adults.

Corvus tasmanicus has only recently been recognised as a separate species and consequently less research on it has been undertaken.

Chapter 9

JACKDAWS

Jackdaws, together with other members of the Crow family, were the first birds to adapt themselves to man in his role of herdsman and later as an agriculturalist. The Romans gave them the name *Monedula*, and the species is now known as *Corvus monedula*. It is not classed as a monotype, although rather distinctive, but grouped with other members of the genus *Corvus*.

In early Britain the bird was known as 'daw', which means a 'simpleton', and by the sixteenth century it was called Jack daw, the word 'Jack' being derived from its call note 'tchack' (or 'jack'); and later the two words became joined. Because of the Jackdaw's thievish ways and talkative nature both 'daw' and 'Jackdaw' were used as words of contempt for people with these characteristics. In actual fact the Jackdaw is an intelligent bird and the increase in its large flocks is proof of its adaptability.

The race *Corvus monedula spermologus* is widespread over the British Isles and western Europe, but in any particular area it may range from being the most common bird to being entirely absent. Its distribution is limited by its narrow choice of nesting sites; the nest must have a solid base, whether it is on a crag, in a hollow tree, or in a building, where there are largish holes above ground level. Natural or quarried cliffs or banks, holes in trees, and tall buildings, be they churches, ruins, or old towers all make good nesting sites. The Jackdaw is notorious for nesting in chimneys which provokes householders to shoot them in the spring. But it is an amiable bird and generally tolerated. I have seen these sociable, cheerful birds strutting jauntily in village streets in Anglesey; perching and chattering on the stone walls outside the grounds of Blarney Castle, and fluttering around the square at Tulcea on the Danube delta. They roost on buildings in European towns, but seem to keep clear of English towns and favour the sea-cliffs and parkland where

46 The Jackdaw *Corvus monedula* is widespread throughout the British Isles (*Eric Hosking*)

there are old trees. In Derbyshire the population increased from 1954 to 1964 when they nested deep in rock fissures in quarries where men were working on loose limestone. They associate with Rooks and starlings in fields, and wander among sheep, often perching on their backs to remove ticks, or take their wool to line their nests.

Smaller than the rest of the Crows, the Jackdaw can be distinguished by its grey nape and ear coverts, black plumage shot with blue on the back and head, the bluish-green gloss on the primaries and primary coverts, and the reddish-purple of the secondaries. It is unique in its pale blue eyes, and has a short, strong, black bill, with overhanging hair at the base covering the nostrils. The tail is nearly square, but very slightly rounded. Its dapper appearance and pert alertness distinguish it from the Rook or Crow, and its quick, jaunty step gives it a less sedate air on the ground. In flight it has a smaller, more compact form and faster wing-beats. Its call is an emphatic, spirited, crisp and clear 'tchack-tchack', quite different from that of other Corvids.

Food plays an important part in the Jackdaw's distribution. It is mostly a ground feeder; its food consisting of 71.5% animal matter such as insects and larvae, with occasional nestlings and young birds; and 28% vegetable matter, including fruit and acorns. It favours short grass on cliffs and in fields and gardens, and avoids thickets, bushes, woods with high trees and tall, standing cornfields.

Aerobatics are performed with great skill and gusto by Jackdaws. They play in the wind, fully aware of distance and the local air conditions. They allow the wind to throw them upwards, then, with a casual flap of the wings, turn over, momentarily opening their pinions and then dive. The wind projects the Jackdaw through the air at over eighty miles an hour with the bird never out of control. Whole flocks will take part in aerobatics, sometimes in co-ordinated displays, but usually each bird frolics at its own pleasure. It flies with rapid wing-beats, but it can be seen gliding like a Rook around cliffs. Jackdaws are summoned to flock for a flight away from the roost with the cry 'Kia!' and the signal cry for the flock to return is 'Kiaw!' They have a wide range of thirty to forty notes, many melodious and warbling. They can mimic sounds, including the human voice. In the past many thousands of Jackdaws had their tongues split in the erroneous belief that it made them better talkers.

In Britain some flocks of Jackdaws remain close to their haunts all the year round, but some depart in the autumn to other roosts and return in the spring. On the whole they move more than other British *Corvidae*. In Europe the populations are even more mobile. Numbers arrive on the east coast of Britain from the continent from mid-October to the first week in November, and depart from mid-February to the third week in April. On the south coast of Hampshire arrivals have been noted in early April, and others have reached Fair Isle from April to October. Scandinavian Jackdaws migrate to Suffolk and Scotland in the autumn.

Jackdaws are one of the most sociable of the Passerine birds. They are commonly seen in flocks of dozens or even hundreds, and roost gregariously in woods, plantations and on buildings and cliffs. Their breeding colonies are usually smaller than those of Rooks because of their greater difficulty in finding suitable nest sites. Every Jackdaw in a colony knows each of the others individually. They have a very strict social, or pecking, order. There is a

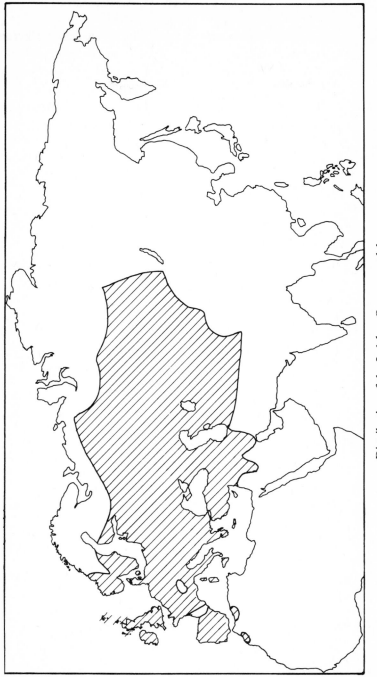

47 Distribution of the Jackdaw *Corvus monedula*

head of the colony and a second-in-command. The head is not aggressive to the underlings but he keeps his direct inferior in place. This degree of animosity to the one directly below the other in rank continues down the flock. The head takes the side of the weaker. In a dispute over rank the rivals draw themselves up to their full height and flatten their feathers signifying they will fly upon each other's backs and claw and fight.

Dr Konrad Lorenz, the German naturalist, who has taught Ravens and Jackdaws to count, explains the snobbery of these socially conscious birds. One of his free flying flock of Jackdaws was a female of low birth, but when a strong male attached himself to the flock and claimed the leadership without opposition, the female made up to him and he succumbed to her charms. After they paired the Jackdaws who had ill-treated her in the past had to respect her and she used every opportunity to snub them.

Jackdaw flocks engage in 'mercy killings' in common with other members of their genus. A Jackdaw injured by gunshot was seen being victimised by a flock and would have been killed had not the observer rescued it.

No wild bird makes a better pet than a Jackdaw. It will show great devotion to its owner and it can be allowed a free run of a garden without causing trouble. It will hunt for grubs and insects; protect itself fearlessly from prowling cats, and, except for picking up any loose grain from the chicken run, it will not interfere with the occupants. One Jackdaw kept tapping on people's windows at night and calling: 'Hello, come on, come on' or 'Who's a naughty boy, then?' He took up residence in a house in Lichfield and had to be removed by the RSPCA inspector, who, realising the bird had been someone's pet, took him home with him.

The corvine fascination for bright objects is strong in the Jackdaw; it is notorious for taking lighted cigarettes and setting a tree or chimney alight. An ex-sailor had one for a pet and it would snatch a lighted cigarette from his lips and fly away with it. An ornithologist found that Jackdaws were first attracted to water by its shine and learnt first to drink, then to bathe and preen and lastly to feed. A shiny insect, leaf or dew-drop would also act as a stimulus for the birds to use their beaks.

In their first year Jackdaws join up in pairs but they do not mature until their second year. Only during the mating period will a Jackdaw provoke his superiors and fight with them to impress his female. He will search for a token nesting cavity and upon finding one will drive all contestants from it with a high, sharp, 'Zick, zick, zick'. In courtship display the male spreads out his feathers and rears up his head and neck. He bows elaborately while giving a lilting rattle of a song, spreading out his tail and pressing his bill to his breast to reveal his grey nape; then he raises his crown feathers, flicks his wings and jerks his tail with a sharp 'yup'. The female squats before her suitor and quivers her wings and tail to signify her submission. When mated the female is self-possessed and aggressive towards the rest of the colony. The pair defend each other loyally and are seldom separated, remaining together during their lifetimes and not even separating during the winter when they join wandering feeding flocks. They become more devoted over the years, continuing to express their affection with loving, low murmuring. The male continues to feed the female during the nesting and brooding periods and the female accepts the food with a low nest-song and later preens her mate's silvery neck while he closes his eyes in ecstasy.

Both sexes build the large nest of sticks. They have a habit of dropping the sticks across a cavity and relying on them to lodge securely. The sticks often pile up several feet high in a large cavity. A nest built in a chimney may reach the fireplace so that the sticks fall out on the hearth; one chimney was cleared of a nest and a wagon-load of sticks was carted away. Sometimes Jackdaws become 'lodgers' in a heron's nest by finding a hole in the side and lining it with wool and hair as they do their ordinary nests. Occasionally they steal a Rook's nest or occupy a Magpie's old nest. They also re-use their own nest-sites. The paired couple defend their nest against intruders with a loud, metallic, rattle note of warning: 'Yiip!' and when they are alarmed on the nest they give a 'Caw', similar, but less deep than, a Rook's call.

During the last days of March when nesting is at its height an aggressive, unpaired Jackdaw may try to oust a nesting couple. The defending male will adopt a threatening posture, making himself look larger by drawing down his head and ruffling his back feathers. The aggressor will adopt the same posture, and both will give a

48 Jackdaw with nestlings (*Eric Hosking*)

short high 'zick'; then they will peck at each other from a distance. Usually the defending male wins, but should the aggressor be the stronger, the nest-owner will give a spate of 'zick, zick, zick' notes gradually changing them to 'yiip, yiip', which brings his mate, her feathers ruffled, to assist him. Should the aggressor still make a stand, the whole colony, with loud 'yipping' will descend upon the nest and with much wing-beating force the interloper to go.

The 4–6 eggs are a unique pale, greenish-blue, spotted and blotched sparingly with black and grey, (35.7 × 25.49mm). The female incubates for seventeen days and is fed by the male. When the young are hatched they have scanty, rather short, pale smokey grey down and, on opening their yellow gapes to beg for food, they reveal purplish-pink mouths. Both parents bring them food in their throat pouches for 30–35 days; the food consists of caterpillars and grassland insects and spiders. The family is very devoted and there is only one brood.

The parents begin to moult when the young are in the nest. The first primaries are moulted gradually and the complete wing moult is not over until the early part of October, although the body moult takes less time. The whole period of moulting takes about 105 days.

The juvenile is browner and less glossy than the adult; the ear-coverts have a silver tinge and the under parts are a paler brown than the upper, with a slight greyish tinge. The tail feathers of both the juvenile and the immature are narrower than in the adult. The juvenile iris is tinged 'pale blue', but at about a month old the fledgling's eye becomes dull brown and gradually turns to dull white, not reaching the pearl-grey stage until the bird is a year old. In its first winter and summer the young Jackdaw is similar to the adult; the juvenile body plumage and most of the wing coverts being moulted in the autumn, but not the wing and tail feathers. The bird is browner and less glossy than the adult and becomes quite brown and dull in the first summer.

The fledged young give a long drawn out 'Karr-r-r' lasting a second, rather like the Rook's 'Caw' but less harsh; this is followed by 'tchack-tchack'. The young have no innate reactions against predators, and recognition of danger has to be learned from personal experience and from their parents. The only instinctive reaction to an enemy is a living creature carrying anything black that flutters, be it alive or inanimate; this will create a furious onslaught on the holder of the object even if it is another Jackdaw, and the birds remain hostile to the holder afterwards. This also applies to Crows, a man with a pet Crow on his shoulder was assaulted by a band of enraged Crows.

The colonial nesting and sociability of Jackdaws enable them to exploit their environment efficiently and reduce mortality among breeding adults, but there is a heavy post-fledging mortality and in the second year the birds suffer from deficiency in food.

The Scandinavian Jackdaw, *Corvus monedula monedula*, occurs in southern parts of Norway, Sweden, Finland and Denmark south to Esbjerg and Haderslev. It has a paler grey 'collar' than the western European race, *Corvus monedula spermologus*, with a small patch of whitish feathers at the base of the sides of the neck, and the under parts are more tinged with grey. It breeds rather later than the Common Jackdaw, during the first week in May in the south of its range and 2–3 weeks later in the north. Its incubation period is slightly longer, ie 18–20 days, and fledging takes 4–5 weeks. It is an autumn migrant. Some Finnish birds winter in Denmark, Sweden, the low Countries and the Soviet Baltic; it has also been seen in Scotland and Suffolk.

Further east *Corvus monedula soemmeringi* ranges from south Russia to east Germany, central Poland, Romania, Yugoslavia, eastwards through the Balkans to the east Mediterranean islands and the Near East; further eastwards it extends from the north-western Himalayas to Kashmir; and in Siberia to about latitude 61°N, eastwards to Yenesei and south to northwestern Mongolia, western and central Altai and to the Sian Shan in Russian and Chinese Turkestan. It is a straggler in Japan. It is a small, slaty black bird which could pass for a miniature House Crow but is distinguished by its pale collar and under parts which are paler than in the typical species; other features are a rather thick neck and greyish white eyes.

This eastern Jackdaw has the characteristics of the Common Jackdaw; it is usually seen in flocks strutting with upright carriage and joining Rooks, Crows and starlings in digging into the turf energetically for food. It keeps company with grazing sheep and cows and follows the plough in search of insects in company with starlings and mynahs. It roosts communally, and nests in small colonies, several pairs sharing a single tree such as a Chenar, in which it occupies the boughs and hollows in the trunks. It also nests in roofs, holes in walls, ruins and cliffs. It breeds in April to June in Kashmir and lays 4–6 eggs (35.1 × 24.8mm), similar in colour to those of the Common Jackdaw.

The juveniles are a dull, dark brown with little gloss and the first year birds are distinguished by their lack of gloss and their hind collar being less distinct than in the adult. In winter enormous flocks commute very long distances daily to and from feeding grounds. In the town of Srinagar, Kashmir, these spectacular flocks are one of the district's great sights during the winter.

In north-eastern Algeria the race *Corvus monedula cirtensis* has grey on its under parts like the Scandinavian form but no white patch on the side of the neck. This Jackdaw is thought to have occurred in Tunisia up to 1880.

The Damian Jackdaw, *Corvus dauuricus*, is a separate species. This bird with a whitish breast and broad white collar extending to the throat, is found in Siberia, Mongolia, Manchuria, China, northern Szechwan and Sikang to eastern Yunnan. It migrates to Russian Turkestan, Korea, Japan, Formosa and China, and south to Fukian in winter. The immature birds resemble a young Common Jackdaw with their indistinct grey nape.

In the words of Konrad Lorenz, who has probably studied Jackdaws more closely than other naturalists, few birds – indeed few of the higher animals – possess so highly-developed a social and family life as the Jackdaws.

Chapter 10

ROOKS

The Rook, *Corvus frugilegus*, has not had an easy existence through-out the ages; its black body and preference for grain have marked it as an agricultural pest. The name 'Rook' is Anglo-Saxon and was derived from the sound of the bird's call.

In the British Isles Rooks and Crows were looked upon as being the same bird until the nineteenth century when more interest was taken in birds, and a clergyman, intrigued by Rooks, declared those in a rookery near his church were tamer on Sunday when man left his gun at home. There is a belief that goes back two centuries that Rooks can smell gunpowder, but their poor sense of smell would scarcely warrant this. The belief that Rooks can spot a man entering a field with a gun is more credible; most Corvids fly away in panic at the sight of a man carrying even a stick, which is an innate reaction against a predator.

After the Napoleonic wars Rooks in London experimented with nesting on buildings instead of on trees and went on doing so until a century ago; they even took up residence on the Tower of London. In 1916 one pair of Rooks returned to breed at the Inner Temple, but they had no success, and by 1945 only one rookery survived in the County of London.

There are few records to indicate how numerous Rooks were centuries ago. In 1944-6 there was a big Rook investigation by the British Trust for Ornithology for the Agricultural Research Council in England, and just over two-thirds of the British Isles were covered under the direction of the ornithologist, James Fisher. The total population of Rooks was assessed to be just under 3 million birds, with some 900,000 nests recorded; a comparison of these figures with those obtained in the 1930s showed an increase of around 20%. The highest density of Rook population was recorded in eastern and southern Scotland, which was the most northern part wholly included in the breeding range. A high density was also found in Ireland. Rooks are not nearly so populous in Europe. In France they breed only as far south as Auvergne and Lyons. They are scarce in south Germany and northern Italy, and are only seen in Switzerland in winter. The species *Corvus frugilegus* stretches southeast to Yugoslavia, Romania, Bulgaria and Russia, ceasing at the Urals. It rarely ventures above 60°N latitude.

The black plumage of the Rook is distinguished from that of the Crow by having a purple, not a greenish gloss, and its head is more red-purple than green. The chin and upper throat are covered with dark, greyish-brown down instead of the Crow's lance-shaped feathers. The adult Rook is distinct from other Corvids by the greyish-white skin of its face, which is covered with short, rounded knobs from its juvenile moult. It has a greyish-black, pointed bill, more slender and less curved than the Crow's. The loose flank feathers on the upper leg give the Rook the appearance of wearing short, baggy trousers. It has black legs and a brown iris. In flight silhouette the Rook has the rounded tail of the Crow, but the third and fourth primaries are the longest and the second is longer than the sixth.

The Rook flies at 27–35mph, which is faster than the Crow. Its flight is direct and deliberate with regular wing beats. It glides freely in casual flights or wheeling over its nest-site, but otherwise only when it is about to alight, or in favourable air-currents. It can be seen flying in loose, straggling flocks calling 'ki-kuck, ki-kook', with a less harsh tone than the Crow. It has around 30–40 varied notes to suit its emotions, all more musical and softer than the Crow's, with a wider range of pitch. It lowers its head as it calls, 'Kaw' or 'Kaak', with its beak, neck and body more or less horizontal and its tail held up in a fan. Its walk is slow and sedate; it keeps its head lowered with an air of deliberation as it watches out for wireworms or cockchafers' grubs disturbed by the plough. It also hops forward or sideways with half-open wings. By gathering in flocks it concentrates in areas where food is abundant.

A flock of Rooks is called a Building of Rooks but the word 'rookery' is used for the communal nesting sites on the tops of trees; it is also used to describe communities of other birds, such as penguins' breeding grounds. Rooks are the most sociable of

birds and their rookeries are the very essence of communal life
where they maintain the same close community through the breeding season as at other times. Their compact colonies may number
thousands of birds. The largest rookery known in the British Isles
is one at Hatton Castle in Aberdeenshire which has contained 9,000
nests. In small flocks of Rooks every bird has a recognised position
in the 'peck order'. Birds of high rank eat before those in lower
positions which ensures the survival of the fittest. There have been
many stories about 'Rook Parliaments'. These consist of some
50–100 birds surrounding one or sometimes two Rooks, and after
much 'cawing' they close in on the victims and peck and beat them
to death. Jackdaws have also been seen to do this and I experienced
much the same thing with my caged Crow. Some say the Rooks are
holding court on a delinquent, but another theory, which I favour, is
that they instinctively set upon a sick, weak, or diseased bird and
perform a ritualised, compassionate killing, thus averting more
disease and putting the bird out of its misery.

As with most of the common Corvids, Rooks make good pets. A
fledgling reared by man will refuse to leave him, even coming indoors with him and following him on his walks. Tame Rooks show
great intelligence; they become adept at letting themselves out of

49 Some Eastern *Corvidae* (*facing page*)
  a *Pica pica bactriana* the White-rumped or Kashmir Magpie of central and
  eastern Russia, Crimea, Astrakan, to the Caucasus, Iran, eastern and northern
  Iraq, Transcaspia, plains of Russian Turkestan, western and central Kirghiz
  Steppes, Afghanistan, Baluchistan, Gilgit, and Baltistan to Ladak. Below:
  showing first primary
  b *Garrulus glandarius leucotis* the Burmese Jay of Burma, Tenasserim,
  Thailand, central Laos, southern Annam and Cochin China
  c *Corvus macrorhynchos culminatus* The Indian Jungle Crow. Below: foot of
  the Jungle Crow
  d *Platysmurus leucopterus* the White-winged Jay of Malay Peninsula, Sumatra
  and Borneo
  e *Corvus frugilegus frugilegus* the Rook of Western Eurasia to Iran, Siberia
  east to Yenesei and northwestern Mongolia, Kirghiz Steppes, Aral Caspian
  region, Bukhara, and the Tian Shan in Russian and Chinese Turkestan
  f *Nucifraga caryocatactes multipunctata* the Larger Spotted Nutcracker of the
  northwestern Himalayas
  (*Illustrations after E. C. Stuart Baker, OBE FZS 1922*)

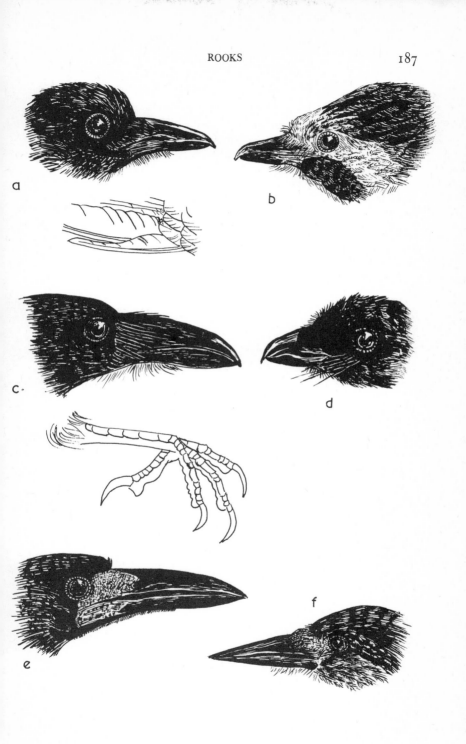

aviaries, learning to unhook the hook and eye fastening on the outside of a cage by pushing the beak through the wire netting. One aviary had a cabin-hook on the inside so that the owner could fasten the door upon entering. The Rook fastened this and the owner had a difficult task unhooking it from the outside!

Flames and heat attract Rooks. One Rook living indoors learnt to switch on the electric fire, and although the switch was three feet away from the fire he was able to associate the two. Another Rook learnt to strike matches with the tip of his beak while holding the match on his perch with his claw. When he had lighted the match he would pick it up, spread his wings and hold it first under one wing and then the other until the flame went out. Rooks enjoy 'anting'; they pick up the ants and place them under their wings and then throw themselves into unusual postures as the ants swarm over them. The most spectacular substitute for ants is fire. One naturalist, Maurice Burton, believes 'fire-anting' may be the answer to the mystery of the Phoenix. He had a Rook named Niger who disported itself with a heap of burning straw. It would stand flapping its wings while flames enveloped the lower part of its body, and smoke drifted around, then it would snatch at the burning embers with its beak and try to put them under its wings. Every now and then it would pose among the flames with its wings outstretched and its head turned to one side in the identical pose of the Phoenix.

In a Rook's roost sentinels are elected to watch out for predators. An example of this occurred when a lady invited a friend to watch Rooks feeding in her garden by peeping round the curtains. The flock of Rooks refused to budge from a clump of elms 200 yards (183m) away, but a solitary Rook, perched in a tree 50 yards (46m) from the house, appeared to be watching the window where the visitor stood. After the friend's departure the Rooks drifted over to the tree where the sentinel perched and soon resumed their feed in the garden.

Rooks search for food by probing the ground with their pointed bills and digging for grubs. They bury food such as acorns and pine-cones and eat them during the winter. In their analysis of 1,577 Rooks' stomachs, the Rook Investigation in England showed that the birds appear to eat 86% of grain and 5% of animal matter, but as grain is digested more slowly than animal matter this was not a correct estimate, because Rooks consume a large amount of insect grubs feeding on the cereals. The investigation showed that six-sevenths of the grain eaten was stubble and much of the rest consisted of ungerminated grains, neither of which are eaten by humans. Given plates of different coloured foods Rooks show a preference for yellow and white, which may be interpreted as their preference for grubs and grain. But a young, tame Rook, when presented with wheat grain, placed the grain in crevices and tried to crush it with its bill, instead of swallowing it whole as wild Rooks do; which suggests Rooks do not instinctively eat grain, but learn to do so by watching their elders. The survey found that Rooks were beneficial to agriculture because they consume large numbers of grubs that feed on the very cereals that the Rooks are accused of destroying. They will not eat germinating grains of corn so they can only feed on the corn for a short period each year. Their most important food for much of the year is earthworms. The rest of their diet is much the same as that of other Corvids.

Rooks do not breed until their third year. In the late summer when the juveniles become independent of their parents they join flocks of adult or wholly juvenile Rooks. From late September to the third week in November the British Rooks tend to make irregular movements. They leave their rookeries and in early autumn take part in aerial displays, tumbling, twisting and diving headlong through the air with 'caws', croaks, grunts and gurgles. They have an unusual habit of hanging upside down from a high tension cable, then dropping, opening the wings, and either flying away or returning for a repeat performance. They have regular morning and evening flights to and from their feeding grounds sometimes thirty miles away from their roosts. They disperse at daybreak and search for food. The adults return to their roosts from time to time and in response to their gradually ripening reproductive organs they sometimes carry out nest repairs and indulge in casual courtship. At night as they return to their roosts, great black hordes can be seen straggling across the autumn sunset. A farmer found a ten-acre field black with some 20,000 Rooks one late autumn afternoon; the next day he put up a Rook-scarer and all the Rooks disappeared. They had eaten nothing and had no doubt gone there to roost. In

the Rook Investigation no proof could be found that cereal crops were more a failure where 300 Rooks were gathered on 1,000 acres as compared with 50 on 1,000 acres. They certainly prefer agricultural country with good clusters of trees and are perhaps the most common farm-bird in the British Isles, feeding off the farms and roosting in the neighbouring trees.

Rooks nest and roost in most trees, but favour elms, oaks, beech and ash; they also use pines, but do not like poplars. They are ground-eaters and spend much time on the ground, but rarely roost, and never nest, there. They are wary of small enclosures and tall grass and crops. In England most rookeries are below 400ft (122m), but in suitable conditions Rooks will breed at over 1,000ft (305m).

By January the male is feeding his female and when their reproductive organs become ripe in February they begin their courtship in real earnest. They communicate in deep churring notes and an occasional peculiar 'burring'. The male flies over his mate with his throat pouch bulging, then struts and postures before her with his head held high, wings drooping and his tail feathers fanned and erect. He will bow to his female several times, cawing loudly. The female responds by crouching and shivering her wings, giving a slight responsive bow or taking the initiative on the ground by lowering her head, trailing her wings and erecting her slightly fanned tail over her back. The male then empties his pouchful of food into her bill. This pouch is an opening under the tongue which bulges enormously when it is filled with food. The mating pair preen and fondle each other's bills. Coition takes place on the nest, and both feeding and coition continue well into the incubation period.

Although Rooks nest in colonies they are very territorial and pairs defend a small area around their nests and threaten and drive off intruders. Old nests are used by past breeders and refortified. The male provides the material for the nest and helps his mate build. Sometimes he takes over completely if the female annoys him by pulling sticks out as fast as she puts them in. When searching

50 Pair of Rooks, monogamous and apparently affectionate, at nest (*Eric Hosking*)

for nest material the male perches precariously on a small branch with half-open wings and his tail fanned out as he reaches for twigs; should he drop one he will fly down and pick it up, and should his perch snap he will hold on to it as he flutters to the ground. The deep, bulky nest is built of sticks, plastered down with earth, and lined with grass, leaves and wool etc. A great deal of noise accompanies the building operations in the rookery, high in the trees. Thieving of sticks, and promiscuous overtures to the hen while her mate is away, cause a great outcry and mobbing of the culprit.

The 3-5 eggs are laid during the latter half of March; the hen usually lays one every day and incubates from the first egg. The light bluish-green or greyish-green eggs are paler than the Crow's, with duller ashy grey and brown spots, (40 × 20mm). In Ireland and on the continent they have been known to be erythritic (of abnormal red shade).

The male feeds the hen during her 16-18 days' incubation, and also the nestlings during the ten days their mother broods them. The dark-skinned nestlings are scantily covered with a short, dark, smokey-grey down upon their spine, upper arms, inner arms and thighs. They open their pale, flesh-coloured gapes and mouths to screech for food, and are soon able to watch for their parents with eager, blue-grey eyes.

The nest houses many small beetles, flies and spiders, which act as food for the nestlings. The Rooks eat the eggs that fail to hatch. Deaths occur in the first third of their existence as nestlings when they are only a fraction of the weight they should attain later. The parents feed their nestlings for a longer period than Magpies and Jackdaws. Higher mortality is suffered among their young because they have to find food from further away, whereas Magpies, Jackdaws and Jays collect food near their nests. The most common disease among the young is a parasitic gapeworm, *Syngamus trachea*, which invades the lungs and trachea and causes such debility by obstructing the lungs, that many young birds die just as they are due to leave the nest. It is doubtful whether more than one nestling in five survives to six months.

In some areas of Britain the Rook population has been cut by half. There has been a decline in nesting without any apparent reason. In Derbyshire a decline was noted between 1944-1966, and

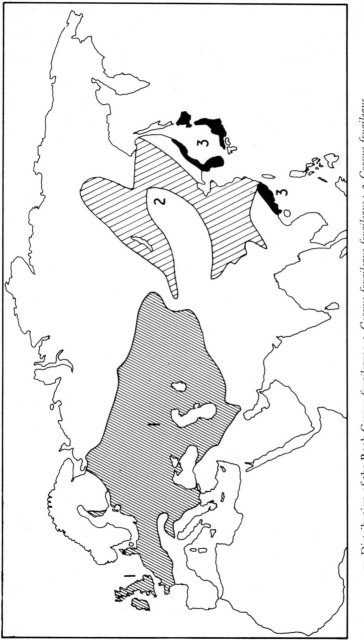

51 Distribution of the Rook *Corvus frugilegus*: 1 *Corvus frugilegus frugilegus*; 2 *Corvus frugilegus pastinator*; 3 *Corvus frugilegus pastinator* winter range

while the Rook population increased in Essex, Herefordshire and Nottinghamshire up to 1958, it decreased in Nottinghamshire after 1962. The decline was probably due to the extensive use of organo-chlorine seed-dressings. Local decreases in southwest Lancashire, Derbyshire and parts of Cheshire have been attributed to a take-over of farmland for industrial development. Two large rookeries have disappeared in the Croydon area, and here none of the known causes for the decline apply. Dutch Elm disease has accounted for many nesting failures because 46% of Rooks favour elms, but they will not nest in dead trees. There are many records of Rooks deserting an apparently sound tree which later falls through being decayed.

In view of widespread reports of the decline of Rook populations in the United Kingdom from the late 1950s to the 1970s, a national census was made in 1975. A preliminary analysis has shown there are 236,769 occupied nests in west England and Wales, 250,000 nests in Scotland and out of a coverage of 49%, 55,022 nests in Northern Ireland. With the data for east England still not complete, a rough estimate of the total British and Irish population is about 1.5 million pairs.

In Sweden there is evidence of a serious decline in Rooks due to mercurial seed-dressings.

Young Rooks become fully grown when they are forty days old, but they are unable to feed themselves until they are sixty days old, and are not really independent of their parents until five months. During the brooding and the fledging period the male may roost near the nest or in a separate tree in the rookery. The cawing of the Rooks continues well into the night and from the close of the nesting season until autumn the birds may still use the rookery roost in small groups, while others link up with different rookeries.

During winter the juvenile loses its facial feathers and black nasal bristles which have given it the appearance of a Crow, and at the end of the first winter the rough grey patch of skin begins to show and gradually the entire face becomes bare. By the next spring the bird has reached maturity and the bare patch is very conspicuous. The rest of the juvenile plumage is not moulted and it has a brownish tinge; by the first summer the plumage is very brown and without much gloss, which distinguishes it from the adult. From

52 Rooks fighting (*Eric Hosking*)

August of the first year the juveniles either flock together in small numbers amongst the adults, or roam for food with wholly juvenile flocks.

During the breeding season the parents moult, beginning with the primaries in May and finishing with the head in September, the whole operation taking about 107 days.

The European Rooks come across the North Sea and the English Channel from Scandinavia and north-central Europe to join up with British birds to scout the fields, strut behind the plough, pick ticks off the backs of sheep and peck among dung hills. At dusk they descend in a black, clattering mass upon a chosen line of trees and after a while their loud cawing is silenced in sleep.

Finnish Rooks winter in Denmark and Sweden; Rooks from the USSR descend upon France and sometimes the British Isles. Polish flocks move to northern Italy, France and the Low Countries. The French Rooks are mostly resident like the British. Ringed Rooks have been reported in the British Isles from Russia, Germany and Holland. By mid-February to the third week in April all the flocks have departed to their breeding quarters.

The Eurasian Rook is a winter visitor or passage migrant into northwest India from Baluchistan, West Pakistan to Kashmir, and the Punjab plains; it occasionally occurs as far east as 76°E. It feeds on grain, tender shoots of gram (chick-pea) and also insects, including the gram caterpillar. In northern Persia, Turkestan and southwest Siberia a subspecies, *C.f. tschuisi* breeds. It is slightly smaller than the nominate species with a smaller, more slender bill. In winter it is a common visitor in the northwest Himalayas and the better cultivated parts of northern West Pakistan.

Further east, *Corvus f. pastinator* breeds in northern Mongolia and Siberia, east of the nominate *C.f. frugilegus*; its range stretches eastwards to the Amur Basin and south to Manchuria and China to the Yangtze River. Inland it ranges as far west as eastern Sinkang. It winters in Korea, Japan and southeastern China. It is slightly smaller than the western Rook and has a feathered chin and lores but a bare forehead.

The adult mortality of Rooks is low because the birds do not breed until their third year and also because of their colonial nesting, and sociable habits. First and second year birds have a high percentage of mortality, the latter probably because of insufficient food. Man is the Rook's chief predator. Rook-shooting is a tradition and provides an exciting sport for the gunman. Birds of prey rarely harm a Rook's nest or young; at their approach an alert is given by the Rook sentinel and the whole flock will mob the predator. Old nests of Rooks are however used by the Red-footed Falcon, (*Falco vespertinus*).

Rooks and the rest of the Crow family figure largely in legends and fables all over the world. The folklore of the Crow and the Raven are confused and it is impossible to dissociate both birds completely. Both foretell death and doom, both are connected with water and there is a belief that both peck out eyes. The Rook is credited with the ability to smell the approach of death, which may be why they desert trees that have died and are about to fall. It was considered lucky if Rooks established a rookery on a landowner's property, but it spelt the death of the heir or the downfall of the family if a rookery was deserted for no apparent reason. Magpies, Jays and Jackdaws have also played a large part in forming the folklore of the world.

BIBLIOGRAPHY

AMADON, D. *The Genera of Corvidae and their relationship* New York City: American Museum of Natural History, No 1251 1944
——*A Preliminary Life History of the Florida Jay* New York City: American Museum of Natural History, No 1252 1944
Amazing World of Nature, Part 2 193 London: Readers Digest, 1969
AMERICAN ORNITHOLOGISTS' UNION *Check List of North American Birds* 5th ed Baltimore: Lord Baltimore Press Inc 1957
Animals Volume 1 Nos 1, 12 and 17 London: Purnell 1965
——Volume 6 Nos 4 and 12 London: Purnell 1965
ARDLEY, N. *How Birds Behave* London: Hamlyn 1972
ARP, W. *Avifauna Venezolana (Corvidae ecology in West Central and Western Venezuela)* 1965
AUSTIN, D. L. *Birds of the World* London: Paul Hamlyn 1968
BAKER, E. C. STUART, OBE *The Fauna of British India, Ceylon and Burma Birds* Vol I London: Taylor & Francis 1922
BENSON, C. W. *Birds of Southern Abyssinia* Ibis 1946
Birds of the World Part I Vol 9 London: IPC Magazines Ltd 1933
Book of British Birds Readers' Digest Drive Publications Ltd 1969
BROWN, J. L. *Social organisation and behaviour of Mexican Jay* Reprint from *Condor* Vol 63 1963
CROZE, H. *Searching Image of Carrion Crows* Berlin & Hamburg: Paul Parey 1970
FISHER, J. *The Shell Bird Book* London: Ebury Press & Michael Joseph 1966
FISHER, J. and PETERSON, R. *The World of Birds* London: Macdonald 1964
GODFREY, W. E. *The Birds of Canada* Ottawa: National Museum of Canada Bull: 203 Biological series No 173 1966
GOODWIN, D. *Behaviour of the Jay* Ibis P 602 1951
HALL, B. P. and MOREAU, R. E. *An Atlas of Speciation in African Passerine Birds* London: Trustees of the British Museum (Natural History) 1970
HANZAK, J. *Birds Eggs and Nests* London: Hamlyn 1972
HARDY, J. W. *Behaviour, Habitat and Relationship of Cyanolyca (Jay)* Kalamazoo: Western Michigan University 1964
——*Habits and Habitats of Certain South American Jays* Contr Sci 165 1–16

——*A Taxonomic Revision of New World Jays* New Mexico: The Condor Vol 71 Las Cruces 1976

HEADLEY, F. W. *Structure and Life of Birds* London: Macmillan 1895

HENDERSON, M. *The Rook Population of a Part of West Cheshire* Tring, Herts: Bird Study Vol 15 Maund & Irvine 1968

HOLYOAK, D. *Breeding Biology of the Corvidae* Tring, Herts: Bird Study Vol 14 No 3 Maund & Irvine 1967

——*Behaviour and Ecology of the Chough and Alpine Chough* Tring, Herts: Bird Study Vol 19 No 4 Maund & Irvine 1972

——*Food of the Rook in Britain* Tring, Herts: Bird Study Vol 19 No 2 Maund & Irvine 1972

——*A Comparative Study of the Food of some British Corvidae* Tring, Herts: Bird Study Vol 15 No 3 Maund & Irvine 1968

——*Territorial and Feeding Behaviour of the Magpie* Tring, Herts: Bird Study Vol 21 No 2 Maund & Irvine 1974

HOLYOAK, D. and RATCLIFFE, D. A. *The Distribution of the Raven in Britain and Ireland* Tring, Herts: Bird Study Vol 15 No 4 Maund & Irvine 1968

IBIS Migrant Jackdaws P 633 1934

INTERNATIONAL WILDLIFE ENCYCLOPEDIA London: Vols 4, 5, 9, 10, 12, 14, 15 BPC Publishing Ltd 1969

JOHNSON, D. W. *The Biosystematics of American Crows* Washington: The University of Washington Press 1961

LACK, D. *Ecological Adaptations for Breeding in Birds* London: Methuen 1968

LAMBA, R. S. *The Nidification of some Common Indian Birds* Part I Bombay Natural History Society 1963

——*The Nidification of some Common Indian Birds* Part II Bombay Natural History Society 1966

LOMAS, P. D. R. *The Decline of the Rook Population in Derbyshire* Tring, Herts: Bird Study Vol 15 No 4 1968

LORENZ, K. *King Solomon's Ring* London: Methuen 1970

MACKWORTH-PRAED, C. W. and GRANT, Capt. C. H. B. *Birds of West Africa* Series 3 Vol 2 London: Longman 1973

——*Birds of East and East Central Africa* Series 3 Vol 3 London: Longman 1973

MARSHALL, A. J. *Biology and Comparative Physiology of Birds* Vols 1 and 2 New York and London: Academic Press 1960

Marvels and Mysteries of the Animal World Part 4 Chap 4 Reader's Digest Ass Inc 1964 New York

MASON, C. W. *Structural Colours in Feathers* I and II J Phys Chem 27 201–251 (1923) 401–447

MORITZ, C. *Current Biography, 1964* New York: H. W. Wilson & Co

NICHOLSON, E. M. *Birds and Man* London: Collins 1951

PEARSON, T. G. *Birds of America* New York: Garden City Books 1936

PETERS, J. L. *Check-List of Birds of the World, Family Corvidae* (Harvard 1940)

PETERSON, R., MOUNTFORD, G. and HOLLOM, P. A. D. *Birds of Britain and Europe* London: Collins 1972

RICHMOND, K. *Krark, The Story of the Carrion Crow* Reading: Barry Shurlock 1973

RIDGEWAY, R. *The Birds of North and Middle America* United States Museum Bulletin 50 (4) 973 (1907)

ROLFE, R. *The Status of the Chough in the British Isles* Tring, Herts: Bird Study Vol 13 No 3 1966

ROWLEY, I. *A Fourth Species of Australian Corvids* Canberra: CSIRO Wildlife Research 1967

——*An Evaluation of Predation of Crows on Young Lambs* Canberra: CSIRO Wildlife Research Vol 14 No 2 1969

——*The Genus Corvus (Aves : Corvidae) in Australia* Canberra: CSIRO Wildlife Research Vol 15 No 5 1970

SALIM ALI and RIPLEY, S. DILLON *Handbook of the Birds of India and Pakistan* Vol 5 London: Oxford University Press 1972

SHUFELDT, R. W. *The Mythology of the Raven* London and New York: Macmillan 1890

TEXT, LIVINGSTONE *Birds of the Eastern Forest* Toronto and Montreal: McClelland & Stewart Ltd 1973

THOMSON, A. L. *A New Dictionary of Birds* London: T. Nelson 1964

VAURIE, C. *Remarks on some Corvidae of Indo-Malaya and the Australian Region* New York: American Museum of Natural History Society 1958

WETMORE, A. *Observations on the Birds of Argentina, Paraguay, Uruguay and Chile* United States Museum Bulletin 133 1926

——*A Revised Classification for the Birds of the World* Smithsonian Institute, Museum Collections 117(4) 1–22 (1951)

WITHERBY, H. F. and JOURDAIN, REV F. C. R., TICEHURST, N. and TUCKER, B. W. *Handbook of British Birds* Vol 1 London: H. F. Witherby Ltd 1948

YAMASHINA YOSHIMARO *Birds in Japan* Tokyo: Tokyo News Service 1961

YOM-TOU, YORAM *Food of Nesting Crows in Northeast Scotland* Tring, Herts: Bird Study Vol 22 No 1 Maund & Irvine Ltd 1975

YEATES, G. K. *The Life of the Rook* Chap 3 Shepperton, Middlesex: Allan 1934

ACKNOWLEDGEMENTS

I am grateful to the following people and organisations who have given me patient and persevering help throughout my work on this book: Oliver L. Austin, Jnr, '*The Auk*', Florida State Museum, Florida; C. W. Benson, Department of Zoology, Cambridge; British Museum (Natural History); British Trust for Ornithology; D. Goodwin, British Museum (Natural History), Tring; Eric Hosking, FRPS, MBOU; Jerome A. Jackson, Mississippi State University, Mississippi; Japan Information Centre, London; The Office of Ornithology, Facultad de Ciencias Naturales Y Museo, Argentina; the Reference and Public Libraries, Bexleyheath, Kent; Dr M. L. Roonval, CSIR, Zoological Survey of India, Jodhpur; Terence Shortt, Royal Ontario Museum, Toronto; Professor William J. L. Sladen, John Hopkins University, Baltimore; the United States Bureau of Sport, Fisheries and Wildlife, Washington, DC; The Wild Bird Hospital Society, London; The Wildlife Management Institute, Washington, DC (Mr J. B. Trefethan); the Wildlife Research Division, Commonwealth, Scientific and Industrial Research Organisation, Canberra City, Australia; John G. Williams, Nairobi, Kenya; the Willowdale Library, Toronto, Canada; the Zoological Society of London.

I am also grateful to my husband who has been an enthusiastic helpmate in my ornithological travels and research.

INDEX